MATHEMATICS AND SOCIAL UTOPIAS IN FRANCE

Olinde Rodrigues and His Times

HISTORY OF MATHEMATICS
Volume 28

MATHEMATICS AND SOCIAL UTOPIAS IN FRANCE

Olinde Rodrigues and His Times

Simon Altmann
Eduardo L. Ortiz
Editors

American Mathematical Society

London Mathematical Society

The portrait of Olinde Rodrigues on the cover, from the *Archives de la famille Pereire*, Paris, is used by the kind permission of Mme. Geraldine Pereire-Henochsberg.

2000 *Mathematics Subject Classification.* Primary 01A55.

For additional information and updates on this book, visit
www.ams.org/bookpages/hmath-28

Library of Congress Cataloging-in-Publication Data

Mathematics and social utopias in France : Olinde Rodrigues and his times / Simon Altmann, Eduardo L. Ortiz, editors.
p. cm. — (History of mathematics, ISSN 0899-2428 ; v. 28)
Includes bibliographical references and index.
ISBN 0-8218-3860-1 (alk. paper)
ISBN 0-8218-4253-6 (soft cover : alk. paper)
1. Rodrigues, Olinde, 1794–1850. 2. Mathematicians—France—History—19th century—Biography. 3. Mathematics—France—History—19th century. 4. Jews—France—Social conditions—19th century. I. Altmann, Simon L., 1924– II. Ortiz, Eduardo L. III. Series.

QA29.R58M38 2005
510′.944′09034—dc22 2005048180

---

10 9 8 7 6 5 4 3 2 1 11 10 09 08 07 06

Contents

Preface

In order to commemorate the sesquicentenary of the death of Olinde Rodrigues, we organized a conference that took place on 1 December 2001 at the Department of Mathematics of Imperial College, London, supported by the London Mathematical Society and the Société Mathématique de France. It then became clear that although much progress had been made in the understanding of the life of Olinde Rodrigues, some work remained to be done. So, instead of publishing the papers as given on that day, we took the unusual step of considering the conference as a programme for further work. Not only did most of the scholars and mathematicians who attended the conference spend a considerable amount of time researching some of the questions that required elucidation, but they also invited some other distinguished mathematicians and historians to produce new contributions. The result is the book that we now present to the readers.

We hope that this publication will draw attention to a period of French history, the first half of the nineteenth century, that was extremely creative, as well as influential in Europe's intellectual and social development. The foundations of modern mathematics were laid, and some of the utopian ideas that flourished at that time, as is not unusual in revolutionary periods, later gave rise to great advances in the position of women in society, in the treatment of workers, and in the understanding of housing, banking, and transport as fundamentals for an equitable society. In all this Olinde Rodrigues took some part, as the pages that follow show.

Numerous acknowledgements are given in the individual chapters and need not be repeated here, but the editors would like to warmly thank the contributors to this book, who over many months had to endure our repeated requests for information and clarification. We would also like to thank the London Mathematical Society and the Société Mathématique de France for their support, without which this work would not have been possible. The editors are very grateful to Professor Peter Herzig and also to Mr. Sergio Plata for kind help in the preparation of the TEX files.

Finally, the editors wish to thank Professor Karen Parshall for her valuable comments on an early draft of this book and for her constant encouragement and useful advice.

Oxford and London, January 2005
Simon L. Altmann and Eduardo L. Ortiz

List of Contributors

Simon Altmann
Brasenose College
University of Oxford
Oxford OX1 4AJ, UK
E-mail address: simon.altmann@bnc.ox.ac.uk

Richard Askey
Department of Mathematics
University of Wisconsin
Wisconsin, USA

Paola Ferruta
Fakultät für Geschichtswissenschaft und Philosophie
Universität Bielefeld
S 3 – 230
Postfach 10 02 91
D-33502 Bielefeld, Germany
E-mail address: paolaferruta@hotmail.com

Ivor Grattan-Guinness
Middlesex University at Enfield
Middlesex EN3 4SF, England
E-mail address: ivor2@mdx.ac.uk

Jeremy Gray
Centre for the History of the Mathematical Sciences
Open University
Milton Keynes
MKK7 6AA, UK
E-mail address: J.J.Gray@open.ac.uk

Eduardo L. Ortiz
Department of Mathematics
Huxley Building
180 Queen's Gate
London SW7 2BZ, UK
E-mail address: e.ortiz@ic.ac.uk

Barrie M. Ratcliffe
Université Laval, Faculté des Lettres
Départment d'histoire
Cité universitaire
Québec, Canada G1K 7P4
E-mail address: `Barrie-M.Ratcliffe@hst.ulaval.ca`
Barrie M. Ratcliffe is now an independent
researcher living in Cambridge, Ontario, Canada.

David Siminovitch
Department of Physics
The University of Lethbridge
Lethbridge, Alberta
Canada T1K 3M4
E-mail address: `isiminovitch@uleth.ca`

Ulrich Tamm
Department of Computer Science
University of Chemnitz
09107 Chemnitz, Germany
and
Department of Mathematics
University of Bielefeld
33501 Bielefeld, Germany
E-mail address: `ulrich.tamm@informatik.tu-chemnitz.de`

CHAPTER 1

Introduction

Simon Altmann and Eduardo L. Ortiz

The study of the life and times of Olinde Rodrigues—mathematician, social reformer, banker—presents interesting historiographic challenges. To start with, there is only one published article wholly dedicated to him, and that was in 1925. Also, much of the information about Rodrigues scattered in publications on history, mathematics, and the social sciences is inaccurate. This is, probably, because in each of the activities in which he engaged, Rodrigues was at the margin of the centres of power, so that he did not leave a school of followers or pupils. As a result, what information we have about him is most often found in marginal archives, in some cases deliberately created to reinforce the historical position of his competitors. Past French historians have paid far more attention to Rodrigues as a social reformer than as a mathematician; and mathematicians, because his work was not always attuned to his own time, have taken decades (if not a century) to appreciate its value. As a social reformer, he is presented sometimes as some sort of a failure, because the mantle of the utopian-socialist movement, which he largely recreated after the death of Henri de Saint-Simon, was taken from his shoulders by the charismatic Prosper Enfantin. The historian here has to move with great prudence, since some of the documents that we can use to reconstruct his life have been left by Enfantin and his followers, and a certain amount of bias has to be discounted.

Rodrigues left no papers or archives, and the only published discussion about him probably produced by an eyewitness is from the French historian Clarisse Coignet in 1883. This one article as well as letters from women to the important Paris periodical *Le Globe* (Rodrigues was a precursor of feminism) paint a picture that in some ways contradicts other material. Even for the analysis of Rodrigues as a mathematician, the discreet veil that covers the study of racial problems in some periods of French history presents problems to the historian. Rodrigues was the scion of a Jewish family of long standing in France. We know that, in principle, he could have had access to either the *École Polytechnique* or the *École Normale*, because Altmann, Siminovitch, and Ratcliffe in Chapter 2 show that some Jews were admitted to these schools, at least until 1813; the problem, however, is why Rodrigues, after his brilliant mathematical schooling, did not pursue an academic career. Some of the authorities from whom we asked information claimed that no racial discrimination ever existed in the French educational system, whereas the fact is that after the Bourbon restoration of 1815 no Jewish man (women did not exist for this purpose) tried to follow an academic career, even as a school teacher. Barrie Ratcliffe presents in Chapter 3 conclusive evidence that there were instructions from the highest levels of the educational bureaucracy that Jews would not be acceptable as teachers: this puts to rest the pious thoughts that it was entirely Rodrigues's own choice to abandon academic mathematics and follow his father's career in banking.

Despite the scarcity of direct documentary evidence on Rodrigues, we present in this book for the first time an authoritative account of Rodrigues's ancestry, of his family, of his family tree, and of his schooling. Chapter 2 summarizes all the information we could obtain on these matters, as well as presenting three hitherto unknown portraits of Rodrigues. A fourth portrait, published only once before in 1925, is also reproduced. Chapter 2 also contains, of course, a brief discussion of Rodrigues's mathematical works, most of which are discussed in more detail in the rest of the book. In order to present a rounded biography of our man, however, Chapter 2 must also deal with Rodrigues's career as a Saint-Simonian social reformer and as a banker. A very useful complete list of all the works by Rodrigues, both mathematical and otherwise, is included in the chapter.

Much of the evidence presented in Chapter 2 comes from derivative sources, often memoirs about contemporaries of Olinde Rodrigues. We have been very fortunate in persuading Barrie Ratcliffe, whose work on the Paris Jewish community in the first half of the nineteenth century is well known, to investigate notarial and other archives in order to obtain as much primary evidence about Rodrigues as possible (again, a problem for the historian, since much archival material was burned during the upheavals in Paris during that period). This Ratcliffe has done, and the result is the splendid Chapter 3, which contains much information about Rodrigues and his life not available in any form until now, providing as well a very useful picture of the Jewish community in Paris at that time. Likewise, Paola Ferruta has given in Chapter 4 another powerful insight into Rodrigues's personality and life through the testimony of his wife Euphrasie, of whom some letters are extant in the archives. Ferruta discusses these in full, as well as two letters from Saint-Simonian colleagues of Rodrigues. These letters illustrate in a dramatic way the stresses and strains to which Olinde was subjected within the Saint-Simonian community that he himself had helped create. These first few chapters provide an insight into the significant effect of discriminatory barriers on the European culture of the period.

The education of Rodrigues, as shown in Chapter 2, was somewhat unorthodox, since after finishing his *Lycée* he moved straightaway onto doctoral work, publishing at the same time six important mathematical papers in, of all places, a journal mainly devoted to the work of the *Polytechniciens*, although he was not registered at that school. Ivor Grattan-Guinness, in Chapter 5, clears up this obscure problem and provides us with a useful background on Rodrigues's education and early research in mathematics.

The next four chapters are devoted to a detailed discussion of Rodrigues's most important mathematical works. One of the few pieces of work for which Rodrigues was fully credited in the mathematical literature was his formula for the generation of the Legendre polynomials which was part of his doctoral thesis. This work is discussed at length in Chapter 6 by Richard Askey. In this article, Askey discusses connections between two apparently unrelated results obtained by Rodrigues. The first one is the well-known Rodrigues formula for Legendre polynomials in terms of weighted derivatives of powers of the binomial (x^2-1), which he had derived in his doctoral thesis. The second result is the generating function for the $n!$ permutations of $\{1, 2, \ldots, n\}$, the coefficients of which are the number of permutations of the same sequence with k-inversions, as discussed in a paper that Rodrigues published in 1839. Askey, displaying his well-known mathematical virtuosity, introduces in the Rodrigues formula operators more general than differentiation, as well as generalizations of the binomial theorem, thereby establishing new connections between the two results of Rodrigues.

In the late 1830s Rodrigues engaged in some important work on combinatorics, resulting in four papers that are discussed in Chapter 7 by Ulrich Tamm. In 1838 Rodrigues participated in a discussion of some combinatorial problems initiated by Terquem in the *Journal de Liouville*, and Tamm shows that Catalan's contribution to this work has been overestimated with respect to that of Rodrigues. In a paper of 1839, on a subject also first discussed in the *Journal de Liouville* by Terquem, Rodrigues gave the generating function for the enumeration of permutations with a given number of inversions. As often happens with Rodrigues's work, the formula he produced here is frequently wrongly attributed to other mathematicians. In another paper of 1843 Rodrigues gave a very useful approximation for the central binomial coefficient, also discussed by Tamm, who ends his article with an application of the Rodrigues formula for orthogonal polynomials in the enumeration of alternating-sign matrices.

In 1840 Rodrigues produced a remarkable paper on rotations and their composition. This is extraordinary in two ways: it introduced for the first time a parametrization of rotations not by their full angles but rather by their half angles. This was a point that was missed by some of the greatest mathematical minds until the end of the century and whose importance was not properly understood until the advent of quantum mechanics in the 1920s. The second point about which this paper is pioneering is the implicit introduction of two important symmetry groups, the rotation group and the Euclidean group. Although a few references to this paper appeared in the literature, it was not until Jeremy Gray brought it to the attention of mathematical historians in 1980 that its import was fully realized. We are fortunate in having him present an update of his paper in Chapter 8.

The last chapter of the book, by Eduardo L. Ortiz, gives an account of the fortunes during the second half of the nineteenth century of Rodrigues's work on rotations of a sphere with fixed centre. It is now well known that this work is profoundly related to the concept of quaternions, introduced by Hamilton three years after the publication of Olinde's paper. Ortiz follows very carefully the slow development of the acceptance of quaternions in France during that period and produces good evidence that, even when they were incorporated within the mathematical corpus used in France, the work of Rodrigues was largely ignored until 1901.

CHAPTER 2

Olinde Rodrigues and His Times

Simon Altmann, David Siminovitch, and Barrie M. Ratcliffe

The political upheavals in France from the French Revolution until the middle of the nineteenth century led to profound social and intellectual changes, sometimes reversed by later events—and whatever happened in France was propagated throughout continental Europe. The French Revolution opened education to all and emancipated the French Jews; even after the Bourbon restoration of 1815 much of what had been gained still survived. Likewise, the new social structure encouraged new ideas about the organization of communities, about economics and banking, and about the status of workers and of women. No figure of the period is more representative of all these new forces than Olinde Rodrigues (1794–1851): he exerted an active influence in the social and political life of his period and, as one of the first Jewish mathematicians of the century (a title often given to the German Carl Gustav Jacob Jacobi, 1804–1851, although they were both preceded by the Frenchman Olry Terquem, 1782–1862), he was himself an exemplar of the cultural sea-change of his times. Yet, no other major figure of the period has been so much ignored or misrepresented. The seesaw of events in his own life meant that he never had a centre of power that he could call his own, and as a result he was seen by his contemporaries and by later historians through partial and much distorted views, although in his own time he had been recognized as an intellectual leader who exercised considerable influence on his colleagues, mathematicians and social utopians alike. In many biographical entries he is only described as a Saint-Simonian or as a social reformer, totally ignoring the fact that he was a businessman whose economic ideas were influential in the development of the French economy. What is even more remarkable, in the midst of all these activities he produced some mathematical work that was so advanced that its real importance was not recognized until the end of his century—to be again forgotten to the extent that his very name was traduced in the mathematical literature as that of two separate mathematicians, *Olinde* and *Rodrigues*. Even the date of his death was often and inexplicably misrepresented.

No single biographical paper on Rodrigues has been published since (Courteault 1925). Because of this lack of biographical knowledge, we shall try to put together whatever information on his life that we could obtain either from contemporary or from secondary sources, especially historical works on Saint-Simonians and on the Pereire family, whose archives are extant and, since they were close relatives of the Rodrigueses, refer to him from time to time. Given the need to fill in so many gaps in our knowledge of this remarkable man, we shall proceed systematically, first dealing with Olinde's origins and family.

Family

Rodrigues Olinde@Rodrigues, Olinde (1794–1851) Ferdinand and Isabella, as is well known, expelled the Jews from Spain in 1492, and Emmanuel did the same in Portugal in 1496. From the time of the transfer of Bordeaux from English to French suzerainty in 1453, Marrano Jews from the peninsula, persecuted despite their formal conversion to Christianity, began their flight to that town, then the most important French port and an active centre of commerce.[1] After a couple of centuries, many of the Marranos there had openly returned to Judaism and, despite some restrictions placed on the Bordelais Jewish community in 1718, it continued to prosper. It is to Bordeaux that Olinde Rodrigues's great-grandfather, Isaac Rodrigues-Henriques, fled.

Information about Olinde Rodrigues's life and family has until now been not only incomplete but often incorrect. For this reason we now present complete evidence about his family, and we hope to clear up the uncertainties that have led to contradictory or erroneous information in the literature. At the same time, we shall provide a great deal of new genealogical material. Olinde Rodrigues's great-grandfather, Isaac Rodrigues-Henriques, was born in Spain, and emigrated to Bordeaux probably in the first quarter of the eighteenth century. Fortunately, excellent work has been done on the Bordelais Jews, and we have been able to reconstruct most of the Rodrigueses' family tree. (See Figures 1 and 2. In order to simplify the presentation of these figures, we do not include children who died in their infancy.)[2] Further details about Olinde Rodrigues's siblings, as well as about his children, until now unknown, are shown in Figure 2.[3]

In the study of the Rodrigues-Henriques family, difficulties often arise because of the frequent repetitions of given names, which can create confusion. Even more, there is another large and distinguished family by the same surname which had long been thought to be entirely unrelated, although some cross-links between both families exist.[4] It is now believed that both Rodrigues-Henriques families have a common root, Moÿse Rodrigues-Henriques, who married Rachel Mendès Campos, and whose children include both Isaac 'La Poudrayre' and Abraham Rodrigues-Henriques (c. 1670–c. 1767). It is only recently that notarial documents have come to light that show Abraham and Isaac, who were believed to be unrelated (probably because of a family feud that kept them apart), to be brothers. Abraham married Rachel Fernandes, and their issue originated the large Rodrigues-Henriques family now thought to be one of the two branches of the family with this name.[5]

---

[1]The history of the French Jewish community is very well given in (Benbassa 1999) and (Nahon 1989, pp. 46–74).

[2]Information for the larger part of Figure 1 comes from (Cavignac 1987, pp. 111–112, 191, and 167), although the names of Olinde's six sisters are from (Ratcliffe 1972, p. 192) and those of Olinde's aunts have been kindly provided to us by Dr. Patrice Assouad. We differ from Cavignac in the date of death of Olinde Rodrigues, which he gives as 26 December 1850, an erroneous date albeit often quoted in the literature. We quote in the family tree, instead, the date 17 December 1851, as registered in the archives of the cemetery of *Père-Lachaise*, Paris, where Olinde is buried.

[3]Constructed from the archival sources listed below as well as the files kindly provided to us by Dr. Patrice Assouad (a genealogist in Paris studying the Rodrigues family, whose wife is a direct descendant of Olinde Rodrigues).

[4]Gustave d'Eichthalfor instance, a close friend of Olinde Rodrigues, married Felicité Cécile Rodrigues-Henriques, of the Abraham Rodrigues-Henriques family.

[5]We are grateful to M. Paul M. Siméon for this information. More details may now be found from the website of the Rodrigues-Henriques family: www.nebuleuse-rh.org.

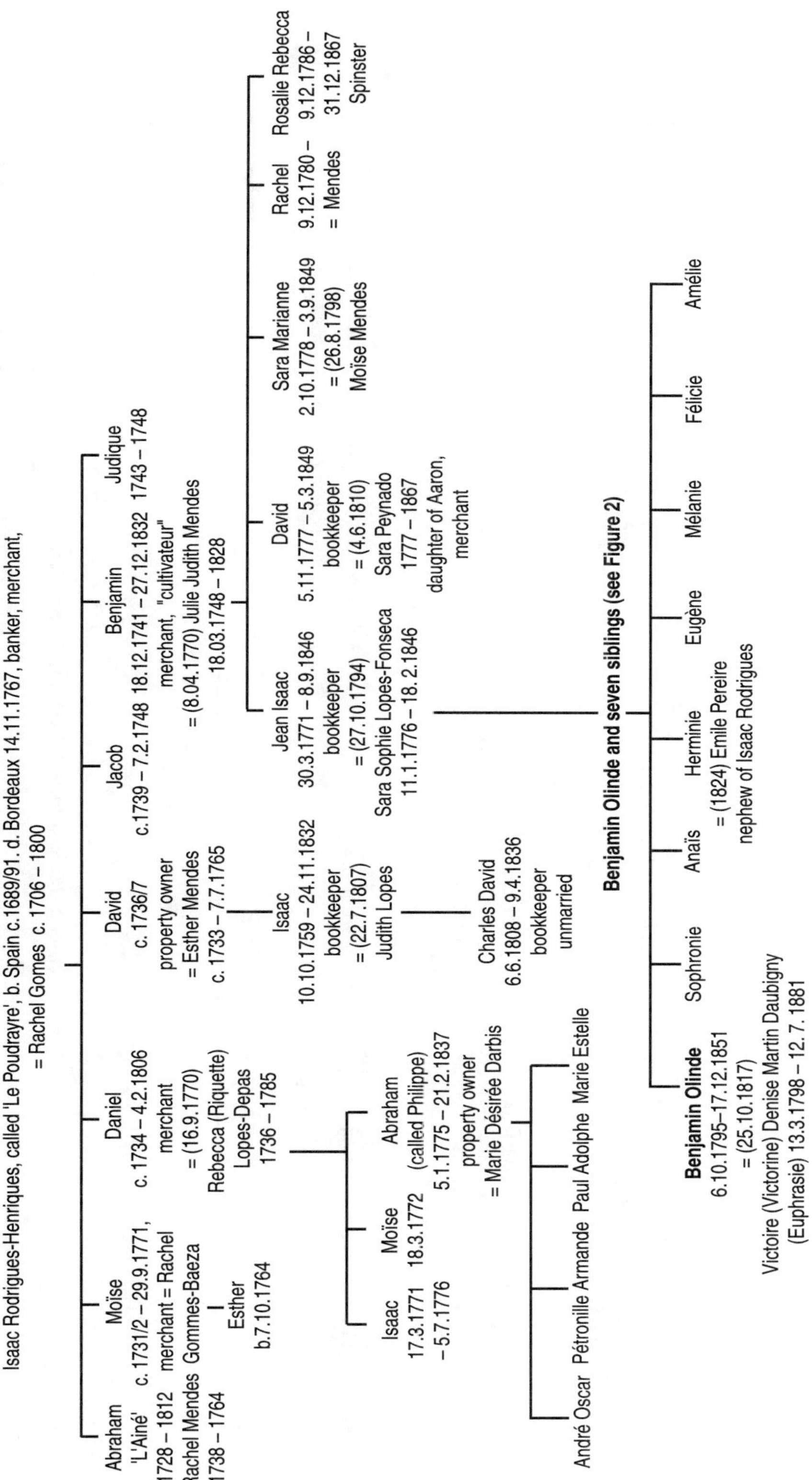

Figure 1. The family tree of the Rodrigues family until Olinde's generation.

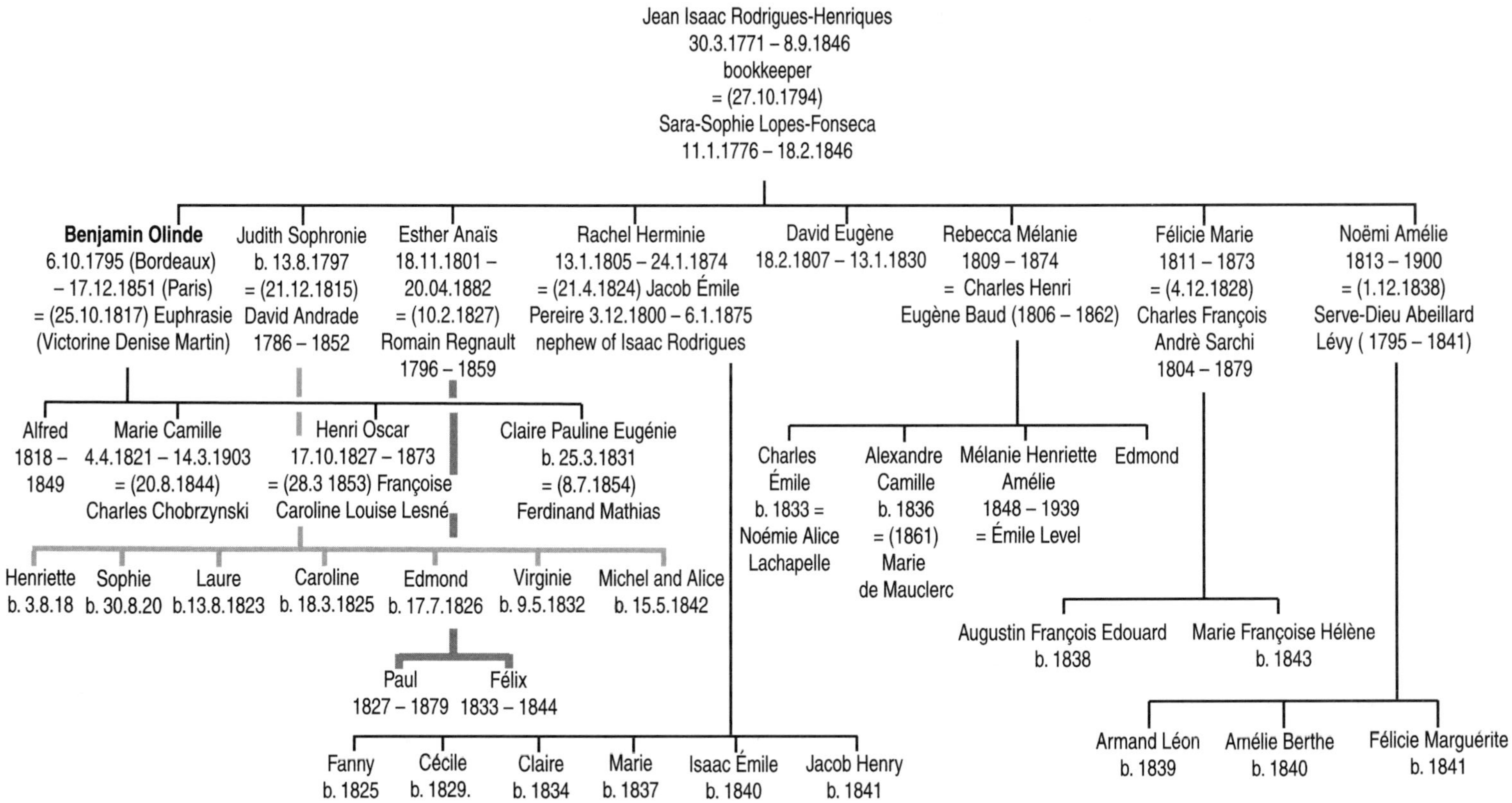

Figure 2. Olinde Rodrigues's family and its descendants.

At about the time of Olinde Rodrigues's birth on 6 October 1795 the Sephardic community in Bordeaux (that is, that composed of Jews of Spanish or Portuguese origin) contained three notable families, the Gradis, the Raba, and the Rodrigues families, and, as with all such communities, was closely knit through intermarriage. (For the Gradis family, see footnote 51.) Another important family, the Pereires, would later be closely connected to the Rodrigueses in Paris: Émile Pereire (Jacob Émile Pereire, 1800–1875) was both a cousin (his mother Rebecca, married to Isaac Rodrigues-Pereire, was a sister of Olinde's mother) and a brother-in-law of Olinde, as can be seen in the family tree. A word or two has to be said about surnames. Rodrigues is a name common in Portugal, whereas in Spain the spelling Rodriguez is the favoured one. Probably because of this, the Spanish *Enciclopedia Ilustrada Espasa-Calpe* claims a Portuguese origin for the family, but we have found no evidence for this. In discussing such an origin for the Rodrigueses, it must be appreciated that at that time in France '*Portugais*' was used as a synonym for Sephardic.[6] In any case, the French were given to modifying name spellings, *Rodriguès* being a form sometimes used for Olinde's family. Also, the Pereires, for example, must have been originally Pereira, and transformed into Pereire in France, where this spelling is still used, although Péreire is also common for this family. Finally, the 'Henriques' of Rodrigues's great-grandfather's name was soon lost, and Olinde himself never used his given name, Benjamin. In fact, the name 'Olinde' was assumed by him at a later stage, since it does not appear on his birth certificate.[7] Regulations were established in 1807 requiring Jews to modify their family names to avoid confusion, owing to many people sharing the same patronymic (Halévy 1863, p. 6). Soon after, it was required of Jews for similar reasons to add to their given names others of French origin. The Napoleonic decree of 20 March 1808, usually called the Decree of Bayonne, required all French Jews to adopt fixed and definitive first and family names and to register them between September and December 1808. We know for certain that after this time Benjamin Rodrigues became Olinde Rodrigues or Benjamin-Olinde Rodrigues: in the *Registre de Consistoire de Paris* of 1809/1810 (now in New York) our man is registered under the single first name of *Olinde*.[8] In any case, assumed names were not unusual at that time: his wife, Euphrasie, (who was not Jewish and thus was not affected by the decree of Bayonne) was in fact called Victorine Denise Martin on their marriage certificate in 1817, in which Rodrigues himself is named as Benjamin-Olinde.[9] Olinde, of course, is not a common name, but it is clear that Isaac Rodrigues was a well-read man who decided to attach to his children non-Jewish second names that would not coincide with those of the Saint's Calendar. Thus *Olinde*, *Sophronie*, and *Herminie* (the latter two being the names of two of Olinde's sisters) are characters in *Gerusalemme Liberata* by Torquato Tasso (1544–1595), the episode referring to Olinde and Sophronie having been translated (as always with Gallicized Italian names) by Jean-Jacques Rousseau in 1781. Likewise, the assumed names of Olinde's other sisters come from literary or musical sources.[10] It is not unlikely that, faced with the need to find

[6]See (Cavignac 1991, p. 279). Relations of the Rodrigues family, the Pereires, hail however from Portugal: Jean-Abraham Rodrigues Pereira (1674–1735) was born in Chacim.

[7]Olinde's birth certificate is given in full in (Courteault 1925, p. 152).

[8]We owe this information to Mme. Stéphane Toublanc.

[9]See Chapter 4 by Paola Ferruta in this volume.

[10]We owe this information to the vast set of genealogical notes about the Rodrigues and Pereire families generously made available to us by Dr. Patrice Assouad. The names *Olinde,*

non-Jewish names for the registration of 1809, Isaac Rodrigues made a wholesale use of such literary and artistic sources as he had at hand.

The profession of bookkeeper (*teneur de livres*) which following Cavignac is quoted for Jean Isaac Rodrigues-Henriques, Olinde's father, in the family tree might not reflect the true status of this interesting man (Ratcliffe 1972, p. 193). On Olinde's birth certificate (Courteault 1925) he is described as merchant (*négociant*). It must be appreciated that at that time and in some financial establishments, bookkeepers were not mere accountants but, acting also as cashiers, had considerable financial clout in the firms for which they worked. In any case, although Isaac Rodrigues worked in Bordeaux for fifteen years as a bookkeeper for the firm of André Aquart, when he moved to Paris later in the 1790s, he worked as exchange agent for the banker Fould (Autin 1984, p. 22) and published a treatise and a manual on accountancy in 1804 and 1810, which include work on banking. Later, he became an independent stockbroker (*kerb-broker*), although outside the Paris Bourse. Isaac Rodrigues had a major role in the Sephardic community of Bordeaux and indeed of France as a whole. After the Revolution, the Bordeaux Jews petitioned the States-General in Paris for emancipation and citizenship, which was granted in 1790 and extended to all French Jews in 1791. Napoleon, who felt that his prejudices against the Jews required revision, summoned an assembly of Jewish notables in 1806, in which Isaac Rodrigues participated as lay member for the Seine department.[11] This *Grand Sanhedrin*, as it was called, proved to Napoleon's satisfaction that Judaism and loyalty to the state were not incompatible, a result for which Isaac's contributions were significant. Although not really well-off, he was certainly not poor: the dowries of his six daughters amounted to a total of 37,575 francs, but Olinde received only 1,500 francs on his marriage (Ratcliffe 1972, p. 193). To compare, during the First Empire a house in Bordeaux could be bought for around 35,000 francs. The Rabas, the richest family of Bordeaux, on the other hand, were worth some 2,000,000 francs in 1827. (See Cavignac 1991, pp. 185 and 253.)

It must be stressed that Isaac Rodrigues must have had a significant influence in Olinde's life in providing him with close contacts with the Parisian Jewish community, both in financial circles, such as his relatives the Pereires, and intellectual circles, such as the Halévy family, who counted composers and writers amongst its members. He moved in fact, probably around 1830, to an apartment on rue Montholon (present *9ème arrondissement*) directly below the Halévy's apartment on the third floor, often visited by the brothers Émile and Isaac Pereire (Halévy 1863, pp. 16–17). Here Olinde also frequented Jacques-François-Fromental-Elias Halévy, best known as the composer of the opera *La Juive*, and who in 1842 married Léonie Rodrigues-Henriques, Olinde's cousin.[12] Olinde had six sisters, one of whom, Rachel Herminie, married a first cousin, Jacob Émile Pereire. Olinde's only

Sophronie, Herminie used by Tasso are stated by editors of his work to be probably 'nomi immaginari' (personal communication from Professor John Woodhouse). Such choice of names shows the degree of acculturation of Olinde's parents. Tasso was popular in France at the end of the eighteenth century through such works as the five-act play *Olinde et Sophronie*, written by Louis-Sébastien Mercier in 1771. The name *Anaïs* for Olinde's sister born in 1801 is that of the principal character in the opera *Anachréon chez Polycrate* by André Modeste Grétry, first performed in Paris in January 1797, only months after the arrival of the family in Paris.

[11]See (Nahon 1989, p. 65) and (Ratcliffe 1972, p. 197).

[12]See (Locke 1986, p. 95). Eventually, Léonie and Fromental Halévy's daughter married Berlioz.

brother Eugène, to whom he was very close, was delicate, suffering from asthma, and died just before he reached the age of 23 (Courteault 1925, p. 156.)

Despite the central role of the Rodrigueses in the Jewish community, they were no longer observant, reflecting a very common attitude within the recently emancipated Jewish élite. Olinde's wife, Euphrasie, was in fact a Roman Catholic.[13] They were very young when they married in 1817 (Olinde, 22, and Euphrasie, 19), and they lived at 26 rue de l'Echiquier.[14] They had four children, two males, Alfred and Oscar, of whom we only know that the second married Françoise Lesné in 1853 and that he was a stockbroker in the Bourse from 1861 to 1873 (Ratcliffe 1972, p. 193.) Olinde's elder daughter, Marie Camille, outlived her father by more than fifty years, and he also had a younger daughter, Claire Pauline Eugénie. Details of their marriages may be seen in the family trees given in this chapter.

Education

Isaac Rodrigues was one of the first Jews to avail himself of the new freedom of sending his sons to the *Lycées* in Paris, where both Olinde and later Eugène became pupils. Olinde enrolled in 1808 in the *Lycée Impérial* (now the *Lycée Louis-Le-Grand* on the Boulevard Saint Michel). We know that at 17, in 1812, the precocious Olinde was already a class assistant (*maître d'études*) in the mathematics course run in the *Lycée Napoléon* by C. Dinet,[15] who had some years before been the teacher of Augustin-Louis Cauchy (1789–1857), later the leading mathematician of his time. Michel Chasles, who became a famous geometer, was one of Olinde's contemporaries at the *Lycée Impérial*, and got the second accessit at the competition at the end of 1811, the first going to Rodrigues, who had already competed the year before (Bertrand 1902, pp. 39–40.) This competition served also for entrance to the *École Polytechnique*, which had been founded in 1794 very much under the influence of Gaspard Monge, great geometer and close collaborator of Napoleon, and where Rodrigues's teacher, Dinet, had been a member of its first cohort.

The school was a high-calibre military establishment, with luminaries such as Arago and Cauchy amongst its first pupils, and Dinet was himself an examiner for admissions to it. Here we have a problem. Bertrand (Bertrand 1902, pp. 39–40) cryptically writes that Olinde 'could not compete except for a prize,' a phrase that could easily be interpreted as signifying that his Judaism prevented him from entering the school. In 1798 Abraham Gabriel Mossé, however, was admitted to the *École Polytechnique*,[16] the first Jew so accepted, and even in 1813 a Jewish friend of Rodrigues, Myrtil Maas, gained admission to it,[17] although he preferred the *École Normale*, which had also accepted him (Ratcliffe 1972, p. 195). So, contrary to claims often made, the *École Polytechnique* was not totally closed to Jews at that time, and it is not clear why Olinde Rodrigues does not appear to

[13]See (Ratcliffe 1972, p. 197) and (Ratcliffe 1971, p. 1231).

[14]Later they lived on 123 Boulevard Pereire and in 1844 on rue Neuve des Mathurins (personal communication from the files of Dr. Jacques Béjot, Paris). At the time of Olinde's death they were at rue d'Amsterdam.

[15]See (d'Allemagne 1930, p. 29). The *Lycée Napoléon* is now the *Lycée Henri IV* at the *place de la Contrescarpe.*

[16]See the students' register in (Marielle 1855, p. 164).

[17]See (Marielle 1855, p. 146). Maas's name is here spelled *Mirtil*, as it also is in the *École Normale* register (Dupuy and Perrot 1895), but the spelling *Myrtil* is frequent in the literature.

have been admitted to it.[18] Pinet (Pinet 1894), in a comprehensive article on the Saint-Simonian *polytechniciens*, mentions Rodrigues only once and certainly not amongst the alumni. It is often stated that he was a pupil at the *École Normale*, but this is false: the exhaustive centennial memoir edited by Dupuy and Perrot gives Maas as a pupil on p. 671 (Vol. 1), but Rodrigues's name never appears in it.[19]

Why, unlike Maas, did Olinde not apply for entrance at the *Polytechnique*? He was already the teaching assistant in Dinet's class at the *Lycée Napoléon*, and it is just possible that Dinet advised him that he was ready to concentrate on mathematical research, directing him for that purpose to the new *Université de Paris* founded in 1808. There is no doubt, however, that he held close connections with the *École Polytechnique*, perhaps through Dinet's influence. In 1813 he taught mathematics to Prosper Enfantin (Booth 1871), who was a pupil there and later played an important role in Olinde's life, but this was probably a private arrangement. It is frequently said that he was a *répétiteur* (tutor) in mathematics at the school, but there appears to be no foundation for this assertion: both Marielle (Marielle 1855) and Hachette (1813) have full lists of *répétiteurs* on which Rodrigues's name never appears. In 1813 Rodrigues obtained his *licence ès-sciences* and the corresponding dissertation was published by the *École Polytechnique*,[20] but such a title could not have been granted by that institution, its degrees being only on technical or military subjects (Pinet 1894). We must therefore assume that immediately after graduating from the *Lycée Napoléon* in 1812, Rodrigues found a home at the *Université de Paris* and that his *licence ès-sciences* had been obtained under the aegis of the *Faculté des Sciences* of Paris, as was certainly the case for his doctorate, also published in the same way.

For about two or three years starting from 1813, Rodrigues was engaged in mathematical research. This must have been so, since in that period he not only obtained his *licence* but from 1814 to 1816 he published six mathematical papers (see the bibliography below). At the same time, from 1814, he was still a mathematics tutor (*maître d'études*) in Dinet's class at the *Lycée Napoléon.*[21] His research was presented to the University of Paris, where he was granted the degree of *docteur ès-sciences* in 1815.[22] The *Faculté des Sciences* required two theses to be presented for the doctorate,[23] although they were not expected to be of a very high standard. Rodrigues's theses were amongst the first presented to the university,

---

[18]The comprehensive lists given by Marielle (Marielle 1855) never mention Rodrigues's name in any capacity. The head of the school's archives, Mme. Claudine Billoux, has kindly confirmed to us that there is no reference whatsoever in them to Rodrigues, although he might have attended classes there purely as an unregistered '*auditeur*'.

[19]See (Dupuy and Perrot 1895). Also, the secretary of the *Association des anciens élèves de l'École Normale Supérieure* has kindly confirmed to us that Rodrigues's name is not given in the list of *normaliens*.

[20]See (Rodrigues 1813), republished in (Hachette 1808, pp. 36–37).

[21]Unpublished notes by Charles Lambert at the *Bibliothèque de l'Arsenal*.

[22]See (Rodrigues 1815*a*). The entry for this publication in our bibliography is quoted from (Ratcliffe 1972, p. 195), where Rodrigues's degree is given as *docteur ès-lettres*, whereas in all his later publications in (Hachette 1808) he is listed as *docteur ès-sciences*.

[23]See (Grattan-Guinness 1990, p. 109). We are grateful to Professor Grattan-Guinness for drawing our attention to this point.

and his first one was published in (Rodrigues 1816*f*), whereas the second part is probably (Rodrigues 1816*a*).[24]

It is not clear who were his mathematical mentors during that period, but the name of Sylvestre-François Lacroix, a former pupil of Gaspard Monge at the *Polytechnique*, is the most likely one. Admittedly, Lacroix was appointed professor of mathematics at the Sorbonne only in 1815, but he is mentioned in a footnote of Rodrigues's thesis both as dean of the faculty and as chairman of the examining committee. This thesis contains the only piece of mathematics for which Rodrigues was ever properly credited, the famous 'Rodrigues formula' for the Legendre polynomials, used and named in this fashion to this day, although until Hermite discovered this work, it was called the Ivory and Jacobi formula.

By the time Rodrigues got his doctorate, the good years for the reformers were coming to an end and with them any possibility of Rodrigues becoming an academic mathematician. After the 1815 Restoration the Catholic hierarchy took control of educational and academic institutions: Gaspard Monge was removed from his post at the *École Polytechnique* that he had helped create and from the Academy as well, and he was replaced by Cauchy, undoubtedly a greater mathematician but with strong Jesuit connections. Whatever Jewish mathematicians there were could not obtain teaching positions, and most of them abandoned the subject or used their mathematical abilities in applied work for private enterprises.[25]

Under these circumstances, it is amazing that Rodrigues retained his interest in mathematics for most of his life. In the first few years around the time of his doctorate he worked in collaboration with his friend Myrtil Maas, who had done very well as a mathematician in the *École Normale* but after graduation had to accept the fact that academic jobs were in practice closed to Jews. In 1818 Maas returned to Paris, where he had employment in the recently founded *Compagnie Générale des Assurances sur la Vie et Contre l'Incendie.* He early formed a committee, of which Rodrigues was a member, to control the actuarial side of the company, and the principles that they established were soon adopted by all the French life insurance companies. Maas and Rodrigues created the statistical tables for the *Union* company of which Maas was the first director (Ratcliffe 1972, p. 195). In 1818–1820 he and Maas also used their mathematical training to draw up the actuarial tables that would be used by all French insurance companies through the century, tables that would eventually be published in 1860 and—another sign of their continuing value—that would go through no fewer than seven editions up until 1933 (Pereire 1860).

Despite the frustration of any ambitions he might have had to become an academic mathematician, Rodrigues did not allow his considerable ability to go unused. We shall first sketch his mathematical work, a backbone to his life.

Olinde Rodrigues: The mathematician

It should be quite clear by now why in modern times there were no Jewish mathematicians in Europe until the beginning of the nineteenth century. Although Rodrigues was first, Jacobi (1804–1851) in Germany was more fortunate in being able to become a professional mathematician, and he thus occupies a more central

[24]See the article by Grattan-Guinness in Chapter 5.

[25]For evidence of the official impediments for Jews to enter the academic world in this period, see Chapter 3 by Barrie Ratcliffe.

place in the mathematics of the first half of the century. This was a crucial time in the history of mathematics, Gauss, Euler, D'Alembert, Cauchy, and other luminaries having prepared the ground for the creation of modern mathematics.[26] The importance that the subject acquired is shown by the founding of the first two purely mathematical journals in the world. An old pupil of the *Polytechnique*, Joseph-Diaz Gergonne, founded in 1810 the *Annales de Mathématiques Pures et Appliquées*, but this was soon discontinued and was revived as the *Journal de Mathématiques Pures et Appliquées* by Joseph Liouville in 1836, an important publication most often cited as *Journal de Liouville.* August Leopold Crelle, in Germany, founded in 1826 the *Journal für die reine und angewandte Mathematik*, also normally cited as the *Crelle Journal.*

Rodrigues's contributions were somewhat scattered across time because of the nature of his extensive activities outside mathematics and can roughly be grouped into three clusters,[27] the first of which consists of nine publications. The first two works listed in our bibliography are his licence and doctoral theses, respectively, later published as we shall describe, and the third, on double integrals and radii of curvature, is related to the thesis for Rodrigues's licence. The next six works were all published in the *Correspondance sur l'École Impériale Polytechnique*,[28] although this publication was mainly devoted to the work of pupils and teachers of the school, with which Rodrigues, as we have seen, did not have any formal affiliation. Nevertheless, 'remarkable' pieces of work from external authors could also be published there (Billoux 2001). The first of these six works, (Rodrigues 1816*a*), is interesting mainly as a forerunner of his major 1840 paper, and the second, (Rodrigues 1816*b*), was his dissertation for the *license ès-sciences*, (Rodrigues 1813), in which he studied lines of curvature, extending the work of Monge and of his student Charles Dupin. Rodrigues was able to obtain a formula to relate the length of a radius of curvature and the coordinates of the centre of curvature which Gaston Darboux, the great French differential geometer, remarked some sixty years later was fundamental in the theory of lines of curvature.

The next two memoirs (Rodrigues 1816*c*, 1816*d*) are concerned with the equations of motion for a particle, a central problem in mathematical physics, for the treatment of which Lagrange had formulated a fundamental principle, called the principle of least action. Its application, however, was obscure, and the second of these papers showed for the first time how to obtain Lagrange's equations of motion for a particle by very careful application of the calculus of variations, then a fairly new method. Rodrigues's role as the invisible mathematician of the century, alas, prevailed, and this work was ignored. (Remember that he never had any pupils whose careers would have been enhanced by quoting the works of the master.)

---

[26]The mathematical developments in France in this period are very well-reviewed by (Grattan-Guinness 1981, 1990).

[27]See (Royal Society 1871) for a list.

[28]See (Hachette 1808). The contributions that appear here were periodically issued in separate numbers and were later bound; we list these papers by the dates of the corresponding bound volumes, but we also give the dates of the corresponding numbers. (It should be noticed that the third volume in the *Bibliothèque Nationale* is incomplete and contains Number 1 only, the date therefore being given as that of that part, 1814.) The dates of the numbers that we give in the bibliography agree with those used by Professor Grattan-Guinness in Chapter 5, which will help the reader to compare our list with his. We are very grateful to Professor Grattan-Guinness for discussion on this point.

Even as soon as 1837 Siméon-Denis Poisson (1781–1840) (the Poisson of the famous statistical distribution), remarked that '[t]oday, this principle of least action is nothing but a useless rule,' which shows not only his ignorance of Rodrigues's work but also how hard it was for people to understand what Olinde had intuited, that is, the central role of this principle in physics—as later testified by the work of Jacobi and of the famous British mathematical physicist Sir William Rowan Hamilton (1805–1865). No nineteenth-century mathematician, with the later exception of Darboux and of Joseph Bertrand, appears even to have noticed this important work, although soon after the turn of the century Jourdain (Jourdain 1908) thought it sufficiently important, as the title of his book shows, to group it with the work of Lagrange and Jacobi. Years of slumber followed until this mathematical gem was rediscovered independently by the Canadian-born Cambridge mathematician Edward Routh in the 1870s.

These two memoirs are followed by the paper (Rodrigues 1816*e*), in which Olinde consolidates the work of his dissertation on lines of curvature (Rodrigues 1815*b*). They are important papers which follow on and considerably extend the work of Monge and Binet. Anticipating Gauss, he used a spherical mapping of a surface, studied the ratio of the areas of the corresponding surfaces, and arrived at a quantity later called the total Gaussian curvature, which he showed equals the product of the principal curvatures. As another example of the credit Rodrigues deserved going to others, this measure of curvature was credited to Gauss, although it appears not to have been known by the latter. (See (Laptev and Rozenfel'd 1996, pp. 6 and 9).) His doctoral thesis, (Rodrigues 1815*a*), was published in (Rodrigues 1816*f*), and it is the only one of his works for which he did eventually obtain proper credit. In studying the attraction of spherical bodies, one has to use so-called spherical harmonics, relatively new when Rodrigues worked on his thesis (they had been invented by Adrien Legendre in 1784). The core, so to speak, of these functions are certain polynomials, called the *Legendre polynomials.* They were fairly well known in Rodrigues's time, but there were no closed formulae to derive them, and he invented an ingenious way to obtain one such formula, called the *Rodrigues formula*, his only lucky discovery in the sense that his name remained attached to it, not without some trouble, as we have mentioned. (See also Chapter 6 by Richard Askey.) The procedure that Rodrigues devised, based on what he called *generating functions* (which had already been used in statistical problems), was later found most useful in deriving similar formulae for many polynomials of interest in mathematical physics.

We must now jump a quarter of a century to reach Rodrigues's next cluster of five mathematical contributions (1838–1840), a quarter of a century that was nevertheless immensely productive in other ways: amongst other things, he became a man of business, with various degrees of success. Given this large gap of time, it is virtually impossible to guess Rodrigues's motivation for involving himself in this research, although it is quite clear that around 1838 he became interested in combinatorial problems, (Rodrigues 1838*a*, 1838*b*, 1838*c*, and 1839) all being concerned with this subject. By far, the most important of these four papers is the last one. Clearly, Rodrigues was keeping abreast of mathematical research by reading the journals of both Liouville and Crelle, in which two papers appeared in 1838 on a combinatorial problem involving the inversion of permutations. Rodrigues solved the problems treated in these papers by again defining some generating functions, which had been so important in his work on the Legendre polynomials, and it is possible that he was inspired to do this work when he recognized the possibility of their use in this case also. Again, this paper was so assiduously ignored for more than a hundred years that repeated attempts appeared in the literature to solve the problem about which Rodrigues had given the final word. It was not until well into the second half of the twentieth century that this paper was rediscovered by Carlitz (Carlitz 1970).

The next paper that we must now consider is one of the most remarkable pieces of work done by Rodrigues, in which he goes back to the subject of his very first work, rotations, but now of a sphere with fixed centre. The year is 1840, and all we know about him from the historians of the period, who knew nothing about Rodrigues as a mathematician, is that in that year he was 'speculating at the Bourse' (Booth 1871, p. 216), although he was also much concerned about the legislation for the *Banque de France*, as we shall see. The story starts with one of the most prolific mathematicians that ever lived, the Swiss Leonhard Euler (1707–1783). In 1775, when he was already blind, he published in St Petersburg, where he lived most of his life, two papers about rotations. The rotation of a sphere around its centre, for instance, is fully denoted by a line, the axis of rotation, and by an angle about that line, the angle of rotation. Euler proved that two such rotations when performed one after the other produce a third rotation, called their 'product', and as a result he described rotations in terms of three angles, called the *Euler angles*. Rodrigues (Rodrigues 1840*a*) went much further, because, given the axis and angle of each rotation, he produced a geometrical construction that determines the angle and axis of the product rotation. We illustrate this in Figure 3, where the first rotation effected is one with axis $\mathbf{a}$ and angle α, the second with axis $\mathbf{b}$ and angle β. The product rotation is around $\mathbf{c}$ by γ. Of course, it required considerable ingenuity for Rodrigues to have discovered that construction, but we refer the reader to (Altmann 1986, p. 155) for its justification. What is important is to notice that all the angles appear not as such but as *half angles*. As is often true, this apparently minor result is of momentous significance, but this was not realized until the end of Rodrigues's century: even then, it was only when quantum mechanics was developed that the half angle of a rotation became a universally used parameter.

With reference to Rodrigues's geometrical construction, it is worthwhile mentioning that Euler was never near to producing such a figure or even an algebraic analogue: like everybody else for most of the century, he used full angles and never their halves. This figure, very much used in crystallography and frequently discussed in crystallography textbooks, is however universally called the *Euler construction*, and it was, in fact, totally unknown to Rodrigues's contemporaries. Eight years after its publication, in 1848, we find the distinguished Cambridge professor of mathematics G. G. Stokes complaining that he was not aware 'of a geometrical proof anywhere published' of the result that the most general motion of a rigid body about a fixed point was a rotation. He then proceeded to produce a very clumsy discussion based on angular velocities. Accordingly, Sylvester (Sylvester 1850) rose to the challenge to produce a proper first-principles geometrical argument and proceeded independently to produce Rodrigues's construction, although, despite a figure, his reasoning is far less clear than in the work he had ignored.

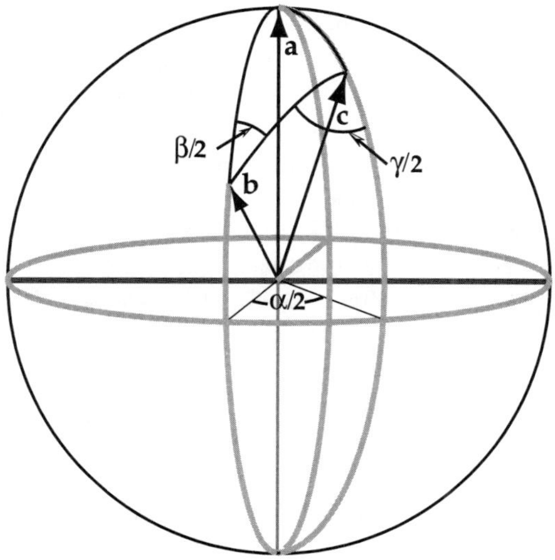

FIGURE 3. The composition or product of two rotations (so-called *Euler construction*).

Rodrigues, however, went a great deal further than his geometrical argument. By using spherical trigonometry in a straightforward manner, he was able to replace each rotation by a set of four algebraic parameters, in such a way that, given the two sets for the two successive rotations, he was able to obtain the same set of four parameters for the product rotation, a truly remarkable result. These are now the essential tool in modern work on rotations, and the four parameters that Rodrigues introduced are universally used in studying spin, in describing molecular structures, in spacecraft kinematics, in robotics, even in ophthalmology, having correctly replaced the use of the awkward Euler angles. The importance of Rodrigues's contribution, however, took some time to be properly recognized. The very successful treatise on the dynamics of rigid bodies by Routh, which reached eight editions, the first one published in 1860, contained no reference to Rodrigues until the third edition (Routh 1877, p. 176), when he proposed that the theorem on the composition of rotations be named after Olinde Rodrigues. Indeed, all the successive editions, up to the last one in 1930, contain a section labelled 'Rodrigues's Theorem.' Such recognition, however, was not readily granted. When the famous German mathematician Felix Klein published his *Vorlesungen über das Ikosaeder* in 1884, he gave Rodrigues some, soon forgotten, partial credit: he grants Rodrigues the discovery of his four parameters and acknowledges that they were unknown to Euler, but he asserts that the latter had used three analogous parameters by replacing the four of Rodrigues by the quotient of his last three by the first. This is a total misrepresentation of the facts, since, as we have said, Euler never used half angles.[29] Every opportunity was taken to bring in Euler and to push the inconvenient Rodrigues away: Schoenflies and Grübler (Schoenflies and Grübler 1902), Klein's collaborators, writing in the most prestigious mathematical encyclopedia of its time, say in relation to these formulae: '[t]hese formulae are quite often treated in connection to Euler. We owe a first essential advance to O. Rodrigues.' The

[29] See the translation of the *Vorlesungen*, (Klein 1956, p. 38).

bible of rotation theory, the monumental work by Klein and Sommerfeld, gives those four parameters without any attribution.[30] Euler's great name of course pushes Rodrigues to second place: to this day the parameters are known as the Euler-Rodrigues parameters. Worse happened to Rodrigues when this work was later discussed: Élie Cartan (Cartan 1938, p. 57), a French mathematician whose contribution to the study of rotations led the world, credits the Rodrigues paper as written by 'Olinde and Rodrigues,' a mistake copied by George Temple (Temple 1960) and repeated by others more than once.

The work of Rodrigues in 1840 is historically even more important, because it should have been used by any unprejudiced mathematician to unravel one of the most extraordinary blunders in the history of mathematics. However, because this blunder was embedded in one of the most important and ingenious creations of the century, people firmly shut their eyes to it. Sir William Rowan Hamilton, the man who in 1865 was to be ranked as the greatest living scientist by the American National Academy of Sciences, had also constructed, three years after Rodrigues, some objects composed of four elements that he named *quaternions* (see (Hamilton 1844)). He obtained them, not from a geometrical point of view, but in an a priori way through algebra. This work, from the algebraic point of view, was outstanding, but from the point of view of rotations, it was an unmitigated disaster that caused a century of confusion and strife. The reason for this is that, having discovered the quaternions algebraically, Hamilton had to interpret them geometrically. He realized of course that his quaternions were related to rotations, but he used two principles that had unfortunate consequences. The first was to give priority to algebra over geometry, whereas it was geometry, which had a strong tradition in his mathematical circle, that had allowed Rodrigues to keep his feet firmly on the ground. The second was to use only full and not half angles, for which he had very sensible reasons, which nevertheless were to be contradicted by the facts a century later. He never read Rodrigues's paper, and he never derived a product rule for the rotations as Olinde had done, but only in a different very indirect way. Cayley (Cayley 1845), who published this latter work before Hamilton did, not only acknowledged Rodrigues but confessed that he did not understand why their convoluted algebraic method gave the same results that Olinde had obtained directly, a question that was cleared up more than a hundred years later by Altmann.[31]

We shall refer the reader interested in the nature of Hamilton's blunder to the careful analysis given by Altmann,[32] which requires a great deal of detailed work, but we shall attempt here to present a rough argument to give an idea of the problem. Using modern vector notation, a quaternion is defined as a pair of a scalar a and a vector, say, $[a, A\mathbf{i} + B\mathbf{j} + C\mathbf{k}]$. A *unit quaternion* may be defined as $[a, A\mathbf{i}]$, with $a^2 + A^2$ unity. Hamilton here fell into two traps. One is that he did not realize that the vector in the quaternion is an axial and not a polar vector, an understandable failure, since he was inventing vectors at the same time that he was doing his quaternion work, and the distinction in question was not known until late in the century. It was this failure, however, that created untold problems when pursuing Hamilton's programme of defining vectors via quaternions. The highly respected mathematician Marcel Riesz (Riesz 1958, p. 21), who first pointed out

[30]See (Klein and Sommerfeld 1897, Vol. 1, p. 56 and Vol. 4, pp. 939–944).
[31]See (Altmann 1986, p. 16) and (Altmann 1989).
[32]See (Altmann 1986, Chapters 1 and 12), (Altmann 1989), and (Altmann 1992, Chapter 2).

this error, qualified Hamilton's interpretation as 'grossly incorrect.' The second trap is that Hamilton noticed that a quaternion $[\cos\alpha, \sin\alpha\mathbf{i}]$ acting on a vector normal to $\mathbf{i}$ 'rotates' it by the angle α and interpreted this as a rotation by α around the axis $\mathbf{i}$ of any vector normal to $\mathbf{i}$. This, however, is not at all a rotation of the vector, but an artifact resulting from the product of two rotations, as was demonstrated by Altmann (Altmann 1989, p. 304). It was clear that there was something wrong in Hamilton's interpretation, since the same so-called 'rotation' acting on a general vector does not transform it into its rotated vector, a result that of course Hamilton knew, but preferred to ignore. In fact, he often took the primary definition of a quaternion as an operator that acting on a vector rotates it into another vector or, what is the same, as the quotient of two vectors. This erroneous statement still appears in the mathematical literature as, for instance, in (Hankins 1980, p. 311) and is the definition that was used by the highly respected Oxford dictionaries until, at Altmann's instigation and with the approval of the chairman of the dictionaries' mathematical committee, Sir Michael Atiyah, it was changed. Even now, it is still given in the current edition of the *Oxford English Dictionary*, although it will soon be changed.[33]

Rodrigues's paper goes much further than we have described. When symmetry operations are combined, their set forms what mathematically is called a *group*. Rodrigues considered not only the rotation but also the translation group, the aggregate of which forms the so-called *Euclidean group*. Translations appear to be totally distinct from rotations, but Rodrigues realized them by producing infinitesimal rotations (rotations by tiny angles) around infinitely distant rotation axes. It was to take more than half a century before this remarkably clever idea was again used. Meanwhile, as is clear by now, Rodrigues's work was largely ignored until its great importance was rediscovered by the mathematical historian Jeremy Gray (Gray 1980), who produced a very penetrating analysis of his article. (See Chapter 8.)

As we have seen, even at this time Rodrigues had not totally abandoned mathematics, and he still read the mathematical literature. He had done nothing along the lines of this paper, however, since 1814, and it is a wonder why he returned to the subject. It must be remembered that geometry had been a strong influence in his mathematical upbringing: Monge had created a powerful school, of which Michel Chasles was an important member. A school friend of Rodrigues, Chasles had continued their acquaintance: we know that they met at the salon of the Saint-Simonians in the 1830s (Bertrand 1902, p. 40), and it is possible that his influence

[33]Hamilton's use of full rather than half angles, as in the quaternion $[\cos\alpha, \sin\alpha\mathbf{i}]$, entails another very important mistake. Given that quaternion form, he took the unit quaternion $\mathbf{I}$, $[0, \mathbf{i}]$ to be a rotation by $\pi/2$ around the axis $\mathbf{i}$. On defining the unit quaternions $\mathbf{J}$ and $\mathbf{K}$ in relation to directions $\mathbf{j}$ and $\mathbf{k}$ orthogonal to $\mathbf{I}$, he then interpreted the relation $\mathbf{IJ} = \mathbf{K}$ as a rotation by $\pi/2$ of $\mathbf{j}$ around the axis $\mathbf{i}$ that transforms $\mathbf{j}$ into $\mathbf{k}$. Although this interpretation is demonstrably wrong (see the correct multiplication rules in (Altmann and Herzig 1994, p. 598), it still is most often stated as the meaning of the quaternion units (see, for instance, (O'Donnell 1983, p. 143)). It can immediately be seen that, on using Rodrigues's parametrization in half angles, the quaternion given should be written as $[\cos(\alpha/2), \sin(\alpha/2)\mathbf{i}]$, so that $\mathbf{I}$, $[0, \mathbf{i}]$ corresponds now to a rotation by π (*binary rotation*) and not by $\pi/2$. Therefore, the equation $\mathbf{IJ} = \mathbf{K}$ really means, following Rodrigues, that a binary rotation around $\mathbf{i}$ times a binary rotation around $\mathbf{j}$ equals a binary rotation around $\mathbf{k}$. This relation is true beyond any possible doubt (see (Altmann 1986, Table **2**-5.2, p. 49)); and it is so fundamental that if it were untrue, the whole of the science of crystallography would collapse.

revived Rodrigues's thoughts on rotations. It is important to remember that precisely at that time, in 1830, Chasles produced one of the fundamental results in mechanics, called to this day the Chasles theorem, namely, that the most general motion of a rigid body is a translation combined with a rotation around a fixed point. It is highly probable, therefore, that this problem was discussed by the two friends and that it remained in Rodrigues's mind.

Three years after the seminal 1840 paper, Rodrigues produced his final cluster of three mathematical papers (Rodrigues 1843*a*, 1843*b* and 1845). The first two are concerned with trigonometrical problems and are followed in 1845 by a short note on continued fractions, his last mathematical paper.

Before we leave this brief discussion of Rodrigues's mathematical works, it is important to mention his writing style. No one who has read the original works of the mathematicians of his period, however great, can help experiencing a sense of relief in reading Rodrigues. His language is concise and entirely free from jargon, and his form of exposition is amazingly clear. Amongst those who have had that experience, the word 'gem' often recurs. The contrast with Hamilton, in particular, is staggering, and one can see why (even (Hankins 1980, p. 322) qualifies Hamilton's *Lectures* as 'unreadable'). Hamilton wrote sometimes starting from preconceptions, and when these proved a trap, he refused to abandon them.[34] Preconceptions were never part of Rodrigues's mathematical writing, which could thus flow from fact to fact without impediment. The opposition of French rationalism with the *Naturphilosophisch* quasimysticism of Hamilton and his followers (it is not without significance that Hamilton was a friend of the poet Samuel Coleridge) could not have a better example than the comparison of these two authors.

Olinde Rodrigues: Banker, social reformer, Saint-Simonian

We have already seen that when it was clear that the Restoration had destroyed any chances he might have had of an academic career, Rodrigues joined forces with Myrtil Maas and worked on producing actuarial tables for insurance companies. The collaboration did not stop there, as their pamphlet (Rodrigues and Maas 1820) on the '*caisses hypothécaires*' (mortgage banks) shows. This exemplifies two interests that were central to Olinde for the rest of his life: financing and how to use it for social purposes. Following his father, he became a freelance broker at the Bourse ('kerb-broker'); soon, in 1823, he was the director of the *Caisse Hypothécaire* at rue Neuve-St-Augustin,[35] and in 1825 he contributed to a pamphlet on banking (Rodrigues 1825). He may have continued for a period to teach mathematics as a private tutor at the *Polytechnique*, but after 1825 there are no further records of any such activities. In 1828 he became managing director of the *Caisse Hypothécaire*, where he would appoint his former pupil Prosper Enfantin general cashier. Probably, his detachment from mathematics in this period is due not only to his banking interests but also to his involvement with the philosopher

[34] As already mentioned, Hamilton first obtained his relation between quaternions and rotations of vectors by considering the rotation of a vector normal to the axis of the quaternion, and although this approach broke down for a vector in a general orientation, he never abandoned his original point of view. (See (Altmann 1986, Chapter 12) and (Altmann 1989, p. 304).) In the latter reference it will be seen, in fact, that Hamilton's original example was not a rotation of the given vector at all, but a product of two rotations that by a freak appears to do what Hamilton claimed.

[35] See (Booth 1871, p. 107) and (d'Allemagne 1930, p. 29).

Saint-Simon and, following the latter's death in 1825, with the movement that came to be known as Saint-Simonism, which was going to be a central part of his life, and the one for which he is best known in the history books on this period.

Claude-Henri de Rouvroy, comte de Saint-Simon, second cousin of the famous Duc de Saint-Simon, was born in 1760, fought alongside Washington in the American war for independence, and not only survived the Terror on his return, but became wealthy through land speculation as a result of it. At the age of 38 he studied mathematical physics at the *École Polytechnique*, Monge, Lagrange, D'Alembert being not only his teachers but also his friends. He had visions of a lay religion, and he proposed at the time the public worship of Sir Isaac Newton. More practically, he devoted himself to numerous proposals for the reorganization of society. A charismatic figure, Saint-Simon inspired his followers with a quasireligious fervour, but his life was very irregular, and he soon went through his money and fell into debt. After the restoration of the Bourbons in 1815 he gradually became more isolated, felt betrayed by his friends, and very systematically, on 9 March 1823, prepared to commit suicide, loading his pistol with seven large-calibre bullets.[36] He then placed his watch on his desk and wrote his last thoughts until the allotted time arrived. His aim, alas, was so poor that of the seven shots to his head only one penetrated his cranium, causing him to lose one eye but not his life (Manuel 1962, p. 113). He was cured in two weeks, although of course he was left in very poor health.

A couple of months later, to alleviate Saint-Simon's illness and destitution, the banker Ardoin, who had already in January introduced his friend Rodrigues to the count, brought him to meet the invalid (also later Léon Halévy, brother of the composer Fromental Halévy) in the hope that the well-off Olinde would provide support.[37] This Rodrigues did and to a very large extent: right from the beginning he took the place of the philosopher Auguste Comte in collaborating with Saint-Simon, and during the next two years until the latter's death he was his constant companion, carefully recording his thoughts. Not only did he support Saint-Simon financially, but he also helped the master finish his last work, *Nouveau Christianisme*, and financed its publication shortly before his death. Thus, he was far more than an amanuensis for Saint-Simon: without him there would never have been a Saint-Simonian movement. On 19 May 1825,[38] Saint-Simon died in Rodrigues's arms, placing on him the mantle of leadership of the Saint-Simonians with the words: '[t]he pear is ripe: you must pick it' (Weill 1894, p. 32). It was then that Saint-Simonism took off as a visionary social philosophy with undertones of religious mysticism. Olinde's brother Eugène, then a brilliant young student at the University of Paris, played a part in proselytizing for the new faith with infectious enthusiasm. He also devised a hierarchy for the membership of the movement, along the lines of a religious community: the leaders, Barthélémy-Prosper Enfantin—who had been Olinde's mathematical pupil when at the *Polytechnique* and later his cashier—Saint-Amand Bazard, and Rodrigues, were called *Pères*, and they in turn addressed the brethren as *Fils*. But it was Rodrigues who provided help, finance, and advice to the movement, and there was much more: for Olinde, as had been true for Saint-Simon, it was important to introduce the aesthetics of romanticism into

[36]See (Weill 1894, p. 30) and (Muso 1999, p. 18).

[37]See (d'Allemagne 1930, p. 29) and (Weill 1894, p. 30).

[38]12 May for (Coignet 1883).

the old Christianity. Later, Enfantin was to say, addressing the Saint-Simonians, '[y]our father Rodrigues was the only one who constantly repeated to us that this book [Saint-Simon's *Nouveau Christianisme*] had enclosed within it the most lofty teaching which was ever given to man to receive' (Manuel 1962, p. 144).

We shall give only that account of the principles of Saint-Simonism[39] that is necessary to understand the circumstances surrounding Rodrigues's life. The New Christianity preached by Saint-Simon was not only an early form of socialism but also a fully fledged universal and lay religion, in which, as noted, even a cult of Sir Isaac Newton was proposed (Muso 1999, p. 86). It must be understood, however, that this was not a mere rationalist approach. The prophet had even said 'God has spoken to me,' and when later a proper cult was formed, a rigid discipline was expected, ruled by the *Pères* and strictly followed by the faithful.

By 1825, with Enfantin's help, Rodrigues founded a journal, *Le Producteur, Journal de l'Industrie, des Sciences et des Beaux Arts*, of which Bazard was also an editor, shares of 1,000 francs each being issued to fund the publication (Charléty 1931, p. 30). It started publication as a weekly on 1 October 1825, but it became a monthly on 1 April 1826, and it ceased publication altogether six months later. Although the paper lasted little more than a year, it helped to unify the group of supporters, and soon the liberal daily *Le Globe*, although more independent of the Saint-Simonians, helped propagate their views, until it was taken over by them in November 1830, when it became the *Journal de la Doctrine de Saint-Simon*. This was largely supported by donations from the Saint-Simonians, although it lasted only until 20 April 1832 (Muso 1999, p. 101.)

The meetings of the community, full of religiosity and excitement, first took place at Rodrigues's *Caisse Hypothécaire* on rue Neuve-St-Augustin, then in salons in Enfantin's apartment at 6 rue Monsigny and later in his house on the outskirts of Paris (now the 20*ème arrondissement*) in Ménilmontant. During this period Rodrigues published three works (Rodrigues 1829, 1831*a*, and 1831*b*), of which the last two are his appeal to the Saint-Simonians at a meeting held in the Salle Tailbout on 27 November 1831 and published in *Le Globe* of the 28th. In it Rodrigues asks all poets to 'sing the hope of a hardworking people which wishes no more to make war.' A new hymn to peace was required to replace the *Marseillaise*, which was a war song. Painters and sculptors were also encouraged to spread the new ideals, and appropriate music had to be forthcoming to surpass the works of Rossini and Beethoven, a proposal to which Mendelssohn took offence (Locke 1986, p. 62).

Enfantin, although he had met Saint-Simon only once (Charléty 1931, p. 65), was not happy as a mere member of the triumvirate ruling the Saint-Simonians and wanted to become their unchallenged leader. His aims, also, were more extreme than those of Rodrigues: not only the founding of a new religion but the very destruction of Christianity.[40] On Christmas 1829 the faithful elected him and Bazard as *Pères* of the *Famille Nouvelle*, and on 31 December Rodrigues abdicated, feeling that his own mission had been accomplished (Courteault 1925, pp. 156–157). His decision might have been influenced by the rapidly declining health of his much loved brother Eugène, who died a few weeks later on 13 January 1830, leaving him disconsolate. This did not prevent him manning the barricades on July 27, 28, and

[39]A concise and clear discussion of Saint-Simonism may be found in (Muso 1999).

[40]See (*École Polytechnique* 1895, Vol. 3, p. 489).

29, when the population of Paris rebelled against the Bourbons and drove King Charles X into exile.

Towards the end of 1831 relations within the community became tense. Enfantin held that he, as a *Père*, was entitled to have sex with the women of the sect (d'Allemagne 1930, p. 217), and he even wrote a letter to Claire Bazard, the wife of Saint-Amand Bazard, as active a leader of the community's women as she was good looking, demanding that she find him a female companion.[41] (Enfantin, in total contrast with Rodrigues, was very charismatic, extremely handsome, and extremely convinced that this was the case.) Not unnaturally, on 11 November 1831 Bazard, and with him nineteen other dissidents, left the community, and very much to his satisfaction, Enfantin became *Père Suprême*. It is surprising that at this stage, despite fundamental disagreements with Enfantin, of which we will say more later, Rodrigues remained loyal to the master, who on 27 November declared a new cult founded that was even more hierarchical. On the next day the Saint-Simonians signed the necessary papers that made them collectively responsible for the group's now perilous finances and gave Olinde power of attorney on all their assets (Charléty 1931, p. 152). In January 1832 Enfantin was confirmed as *Père Suprême*, with Rodrigues as *chef du culte*. This situation did not last long. Admittedly, Rodrigues had become almost a slave of the Bourse, his attendance there being essential for his business, but there were far more serious issues. He strongly objected to Enfantin's outlandish views on sexual morality, which claimed incest and adultery as ingredients of the new moral code. On 12 January 1832, in an article by Duveyrier based on Enfantin's ideas, *Le Globe* announced the impending coming of the *Femme-Messie*, whose arrival was expected to herald a vision of life as a banquet of delights, including free choice of sexual partners.

Already, when early in 1830 Enfantin presented his projected moral code to Rodrigues and Bazard, Olinde qualified this in a letter as a monstrous heresy (d'Allemagne 1930, p. 215). On 17 October 1831 he had presented to the Saint-Simonians counterproposals for a moral code (d'Allemagne 1930, p. 221), but in the power struggle that ensued, Enfantin was not averse to stooping to strike Rodrigues at what the latter held most dear. For Rodrigues, his fifteen-year-old marriage to Euphrasie was central to his life and to his principles, whereas Enfantin relished telling him that his wife had asked him, Enfantin, for help to get away from her husband.[42] This, for all we know, could have been a complete fabrication, Enfantin not being the most veracious of men. All this had become too much for Olinde, and in February 1832 he decided to leave the movement. The break was dolorous: he was expelled from his own apartment on rue Monsigny, for the maintenance of which he had paid 200 francs per annum;[43] in his turn, of course, he blocked all the financial records of the Saint-Simonians (Ratcliffe 1971, pp. 1229–1230). In March, as a final

---

[41]See (d'Allemagne 1930, p. 216). That something more drastic may have happened between the two might be inferred from a later (1832) letter of Claire Bazard to Cécile Fournel, in which she equates Enfantin with Satan (d'Allemagne 1930, p. 239).

[42]See (d'Allemagne 1930, p. 243). This problem is thoroughly discussed by Paola Ferruta in Chapter 4.

[43]Personal communication from the files of Dr. Jacques Béjot, Paris.

attempt to save the movement, Rodrigues proclaimed himself the sole head of the Saint-Simonians, but nothing much came of this.

Their affairs, in any case, were not going well because Louis-Philippe's government started applying pressure on opponents of the regime, and charges were brought in 1832 against several prominent Saint-Simonians, Enfantin and Rodrigues amongst them, on the grounds of outrages against public morality, Enfantin in particular having been entirely open in his support for total sexual freedom. (Moreover, they were accused of illegal association, meetings of more than twenty people being prohibited by Art. 291 of the *Code Pénal.*) The trial was bizarre, Enfantin indulging in exhibitionism, dressed in flamboyant, very lightly coloured clothes, with '*le Père*' embroidered upon his chest. During his deposition he paused from time to time and silently gazed at the judges for minutes, asserting when challenged that 'they would come under the influence exercised by his appearance.... The Attorney General had yet to learn the full power of beauty' (Booth 1871, p. 189). This was all to no avail, since Enfantin was sent to the *Sainte-Pélagie* prison for a year (after which he abandoned public life and became a hard-headed businessman). Rodrigues escaped more lightly, with a nominal fine of 50 francs. (An account of this trial is given in (Rodrigues 1832*b*, 1832*d*).) Three other pamphlets on Saint-Simonism (Rodrigues 1832*a*, 1832*c*, 1832*e*) were also published in this period. The first is particularly interesting and had been published in *Le Globe* on 19 February 1832 as a manifesto on his proposed moral code, which reveals Rodrigues's serious interest in promoting the status of women. This article is in fact a major milestone in the history of feminism.

Despite the schism, Rodrigues remained a Saint-Simonian at heart for the rest of his life: the words told to him by Saint-Simon on his death bed, 'Rodrigues, you must not forget, and remember also that in order to do great things it is necessary to be passionate' (reported in *Le Globe* 30 December 1831 (Charléty 1931, p. 23)) were engraved like a motto in his mind. In 1832, however, he published his last works as a Saint-Simonian, (Rodrigues 1832*e*), and an edition of the works of the master (Rodrigues 1832*f*). Later, he produced two more compilations of Saint-Simon's works: (Rodrigues 1841*b*) and (Rodrigues 1848*a*).

As a banker, Rodrigues understood the importance of securing capital investment and his views, and those of others, contributed to the foundation of new financial institutions, in which the Pereires took a major part (Ratcliffe 1972). A bill to renew the charter of the Bank of France caused Rodrigues to make his own proposals (Rodrigues 1840*b*). Rodrigues held that transport was an engine of social improvement, and some of the first French railways, which played such a pivotal role in France's industrialization, were partly a result of the efforts of several Saint-Simonians. In 1835 a cousin (and brother-in-law) of Rodrigues, Émile Pereire, gained the concession from the government to build the first passenger line, from Paris to St-Germain. Clapeyron and Lamé, old school friends of Rodrigues and marginal Saint-Simonians, were responsible for the construction, Clapeyron himself being in charge of locomotive design (Ratcliffe 1972). In 1846 Enfantin, now settled down as an active entrepreneur (d'Allemagne 1935), became the director of the three railway companies that covered the Paris-Marseille route, as well as secretary general of the Paris-Lyons line. He also formed a group to draw a plan for a Suez Canal and even one for a Channel tunnel.[44]

In 1840 Rodrigues founded a journal, *Le Patriote de 1840*, to address the gulf between the workers and the new but by now well-established bourgeoisie, and he also published a pamphlet on pacifism (Rodrigues 1840*c*). This journal soon closed down, but the contact that Rodrigues had formed with the workers led him to publish a 500-page collection of workers' poetry, *Poésies sociales des ouvriers* (Rodrigues 1841*a*), which was followed by Rodrigues (1841*c*). There was now a hiatus in his public life that lasted seven years, until the 1848 Revolution brought a revival of the Saint-Simonian spirit and inspired Rodrigues to draw up a project for a constitution of the Republic and another for universal suffrage (Rodrigues 1848*b*, 1848*c*). Although of course they were not adopted, they had a significant influence in furthering the rights of both women and workers and in creating an interest in elections. With his Saint-Simonian friend Gustave d'Eichthal (1804–1886), Rodrigues fought for the abolition of slavery, which finally came about in 1848 (Benbassa 1999, p. 121). He even proposed the establishment of working men's holidays, so novel an idea that it was not adopted until 1912. His concerns about the organization of labour are also reflected in (Rodrigues 1848*c*) and (Rodrigues 1848*e*). One of his last works, (Rodrigues 1848*d*), was about banking, and again contained his radical critique of the existing structures of the Bank of France.[45]

Saint-Simonians and Rodrigues himself were not only involved in social and political matters but also exerted a profound influence on the cultural life of France in this period. Although their incursions into music were not always welcome (Mendelssohn resented Olinde's efforts to convert him, as discussed by Locke),[46] they left their mark, for instance on the works of Fromental Halévy. Franz Liszt, who was introduced to Saint-Simonism in 1830, was attracted to the idea of a community based on brotherly love as described in Rodrigues's *Réunion de la famille* (Rodrigues 1831*a*), which he had read with approval and which he commended to George Sand (Locke 1986, p. 101). From his youth Rodrigues had been passionate about music: while he was a *maître d'études* in 1814–1815 at the *Lycée Napoléon*, not only did he teach Charles Lambert mathematics, but he kindled his passion for music.[47] It is probable that this mutual interest was one thing that attracted him to his young bride Euphrasie.[48] He also took an active role in the musical activities of the community, singing lustily at their meetings, and he frequented musicians. Enfantin, for instance, met Mendelssohn at his home (Locke 1986, p. 334).

We do not know much about Rodrigues's last few years except that he of course was much affected by the Revolution of February 1848, which brought him back to libertarian ideas. Also, his social sensitivity was outraged at the high unemployment that the flight of the well-to-do bourgeoisie (and of their capital) had exacerbated.

---

[44]See (Pinet 1894), (Charléty 1931, p. 285), (Walch 1970*a* p. 624), and, especially, (d'Allemagne 1935) and (Walch 1970*b*), where the influence of the Saint-Simonians on industrial developments is fully discussed. Barrie Ratcliffe (1995), however, believes that this influence has been exaggerated.

[45]For lists of these publications by Rodrigues we have consulted (Walch 1967) and (Gerits 1986).

[46]See (Locke 1986, p. 108).

[47]Unpublished notes from the ex-Saint-Simonian Charles Lambert bequeathed to the *Bibliothèque de l'Arsenal.*

[48]See Chapter 4 by Paola Ferruta.

A meeting of old Saint-Simonians was summoned at Rodrigues's house with a view to regenerating the movement, but with the exception of Isaac Pereire, no one agreed to this (Locke 1986, p. 200). Yet, Rodrigues's idealism was not dead, and he went to London, where the Chartists, in sympathy with the Revolution in France, planned a large public demonstration for 10 April 1848. On the eve of that day he was seen near the entrance to Somerset House in the Strand getting up on a chair to harangue the crowd. A policeman told him to hold his tongue and get down, which he refused to do, whereupon the officer pulled him down and hit him with his truncheon (Reid 1902, p. 132). A few years later Rodrigues died in his modest apartment on rue d'Amsterdam (in the present-day *8ème arrondissement*) largely forgotten (Michaud 1843, pp. 288–289). Even the date of his death is often misquoted in the literature: 26 December 1850 in Michaud.[49] The records at the Paris cemetery of *Père-Lachaise*, where he is buried near Saint-Simon's tomb, show unequivocally, however, that he died on Wednesday 17 December 1851, that his wife Euphrasie signed the death certificate, and that he was interred on the 19th: even as a long-lapsed Jew, the Jewish tradition of early burial had been respected. We know that he died in the arms of Isaac Pereire, and it is said that his death was the result of a minor accident (Ratcliffe 1971, p. 1231). On the death of his wife Euphrasie, she was interred in Olinde's tomb, a single headstone commemorating both of them.

As we have seen, history was not very kind to Rodrigues. Most of the short biographical notes on him make no reference to his being a mathematician. Michaud (Michaud 1843) calls him an 'economist and social reformer'; for Courteault (Courteault 1925) he is only a Saint-Simonian. Both Benbassa and Cavignac refer to him, erroneously, as the 'financier of Saint-Simonism.'[50] Booth (Booth 1871) calls him *Rodrigue* throughout his book, as is also the case for Coignet (Coignet 1883), whereas *Rodriques* is the way by which he is known to Edmund Wilson (Wilson 1941, p. 100). Not too long ago, a distinguished professor at the *Collège de France* referred to him as *Olindes* Rodrigues (Chambre 1970). As for the mathematicians, their nominal travesty, as already recorded, is even worse.

Olinde Rodrigues: The man

Many books on Saint-Simonism include pictures of the main protagonists, but we have found none with a portrait or sketch of Rodrigues, even in the monumental and well-illustrated tome of d'Allemagne (d'Allemagne 1930). Until the present time the only picture extant was that given by Courteault (Courteault 1925, p. 157). Fortunately during the preparation of this book three more pictures have been discovered, and they are shown here, together with the one in Courteault's article. Plate 1, which represents Rodrigues as a fairly young man, c. 1830, was found in the Archives of the Pereire family in Paris. Plate 2, the only portrait so far published, appeared in (Courteault 1925, p. 157 and also the frontispiece on p. 152), and it is a reproduction of a portrait that at the time of that publication was in the possession of Mme. Deutsch de la Meurthe.[51] This portrait shows Rodrigues probably in his

[49]See (Michaud 1843, pp. 288–289). This date is also given by Cavignac (Cavignac 1987), although we have not been able to find the origin of this error.

[50]See (Benbassa 1999, p. 198) and (Cavignac 1991, p. 357).

[51]Georgette Deutsch de la Meurthe married (1918) Gaston Gradis, a grandson of Henri Gradis (1823–1905), himself a grandson of Laure Sarah Rodrigues-Henriques, a member of the 'Abraham' branch of this family. M. Henri Gradis, son of Georgette, has kindly informed us that

late thirties or early forties. Plate 3, part of the Pereire family archive, is likely to be a lithograph from around 1840, and it is marked Imp[rimerie] Lemercier & Cie. Plate 4, instead, must be of Rodrigues at a later date. It was discovered by Dr. Paola Ferruta in the *Fonds Eichthal* at the *Bibliothèque de l'Arsenal*, Paris, and is a photograph by the famous photographer Félix Tournachon Nadar (1820–1910), dated 1887 (36 years after Rodrigues's death) and signed 'Nadar'.[52]

As for Rodrigues's physical appearance and character, hardly anything is ever mentioned except for the references in (Coignet 1883). The author of this paper is almost certainly the French historian Clarisse Coignet, born in 1823, who writes as one who knew Rodrigues personally, which is highly probable. Coignet (Coignet 1883, pp. 131–132) describes Rodrigues as small and bony, without being thin, having a regular face with his hair and sideburns curly. His walk was uneven and quick, his talk brief and brusque, sometimes with prophetic accents. As for his character, the evidence appears somewhat contradictory and has to be carefully assessed. Undoubtedly, he suffered in comparison with the charismatic Enfantin: Michel Chevalier compares his authoritarian manner with the 'great magnetizer' that Enfantin was (Ratcliffe 1971, p. 1226). There is an anecdote repeated more than once[53] that makes him appear in the wrong light. In one of the Saint-Simonian meetings it is said that Rodrigues insisted on a testimony from everyone present that the Holy Ghost dwelt within him, Rodrigues, and that when someone expressed disbelief, he fell down in a fit, whereupon a doctor had to be called; but by the time he arrived, Rodrigues was fully recovered, the quarrel having been resolved. This episode, however, has to be interpreted with an understanding of what Saint-Simonian meetings were like. In fact, Coignet (Coignet 1883, p. 159) makes it quite clear that this meeting, in October 1831 on rue Monsigny, came as a result of Enfantin requiring the faithful to submit to total and public confession, no personal trait, however intimate, being beyond discussion. Such meetings lasted sometimes ten hours at a stretch and such was the strain that even young people were subject to fainting during them. Rodrigues, as '*chef du culte*' was acting entirely within Saint-Simonian doctrine when demanding a statement of faith: the priest was supposed to reveal the law and to determine its application by virtue of an afflatus exclusive to him (Coignet 1883, pp. 153–154). Furthermore, if Rodrigues's demand appears excessive, compare it with Enfantin: 'I am Man-God,' stated in a letter to Duveyrier of 21 June 1830 (d'Allemagne 1930, p. 252). Coignet writes that it was the doctor who required a retraction from the doubter, Jean Reynaud, in order to save his patient, who soon recovered, and that such was the pathos of this meeting that a number of other members remained ill for several days (Coignet 1883, p. 159).

---

the present whereabouts of this portrait are unknown. We are grateful to M. Paul M. Siméon for the genealogical information in this note.

[52]This picture must be a photographic reproduction of a daguerreotype or perhaps an engraving done in the 1840s. Fortunately, other known examples exist where records show that Nadar, despite his great fame, used to engage in such practice (personal communication from Gordon Baldwin, of the Getty Museum in California, and Barbara Tannenbaum, Chief Curator of the Akron Art Museum in Ohio). It can safely be assumed, therefore, that this is indeed a reproduction of an earlier picture done in Rodrigues's lifetime. Mme. Françoise Reynaud, Curator of Photography at the *Musée Carnavalet*, Paris, suggests that the original, rather than a daguerreotype, was an engraving, or a lithograph, of which unfortunately there is no trace. Similarities in style and dress with Plate 3 suggest that the original was by the same hand.

[53]See (Weill 1894, p. 102) and (Ratcliffe 1971, p. 1230).

PLATE 1. C. 1830, from the *Archives de la famille Pereire*, Paris. By kind permission of Mme. Geraldine Pereire-Henochsberg.

REVUE PHILOMATHIQUE

OCTOBRE - DÉCEMBRE 1925.

OLINDE RODRIGUES

d'après un portrait de famille.

PLATE 2. Frontispice on p. 152 of the article by P. Courteault (Courteault 1925). Un bordelais Saint-Simonien. *Revue Philomatique de Bordeaux et du Sud-Ouest* **28**, kindly supplied by the *Bibliothèque des Archives Départamentales de la Gironde*, Bordeaux.

PLATE 3. C. 1840, from the *Archives de la famille Pereire*, Paris. By kind permission of Mme. Geraldine Pereire-Henochsberg.

PLATE 4. From *Bibliothèque de l'Arsenal*, Fonds Eichthal, XXXII, manuscript 14110. This photograph is dated 1887 and signed by [Félix Tournachon] Nadar. Reproduced by permission of the *Bibliothèque Nationale de France.*

It is likely that Rodrigues's character, as a man of severe principles and strong opinions, was not an easy one. Claims have been made that Rodrigues did not play a major role in the Saint-Simonian movement because of his irascible character, but such a trait is not mentioned by Coignet. There is a letter from Michel Chevalier to Rodrigues, where he accuses the latter of having the speech and gestures of an 'army commandant,' and this expression is sometimes taken against Olinde. It must be remembered, however, that this is the editor of *Le Globe* with whom Rodrigues had serious doctrinal differences and to whom he addressed an open letter (Rodrigues 1832*c*). Although these words are often quoted, the actual expression used by Chevalier is not quite as harsh: 'commandant of a pacific army of workers,' as can be seen in Chapter 4 by Paola Ferruta, where Chevalier's letter is treated in detail.[54] It must not be forgotten either that the Saint-Simonians were an unruly and argumentative lot and that their behaviour could be totally outrageous—often bound to disturb any sensible person, let alone Rodrigues. That he tried to placate discord is an undoubted fact, acknowledged even by Enfantin at a time when he was no friend of his. Writing from prison, he said: 'I would like to speak of the peaceful sentiment that Rodrigues finally managed to inculcate in us' (d'Allemagne 1930, p. 72). Also, d'Allemagne (d'Allemagne 1930, pp. 124–125) quotes opinions on Rodrigues's insistence on 'peaceful words,' and 'words of peace.' He could be extremely patient, even when dealing with ideas that he did not like: when at a meeting in 1831 Jean Reynaud attacked the morality of Enfantin's ideas on women, he spent one and a half hours calming him down (Locke 1986, p. 91). On the other hand, he was intense in defending the ideas of Saint-Simonism, and he could be tedious in proselytizing for them, which sometimes put people's backs up, as was the case with Mendelssohn.

The most detailed study of Rodrigues's character is given by Coignet (Coignet 1883, pp. 131–145) who shows him as sincerely disinterested personally and totally devoid of egotism. On the other hand, as a religious puritan, he was sometimes prepared to sacrifice individual rights and freedom. 'He spoke little and never from his own authority,' and he was capable of the greatest sacrifices. During the debates at the end of 1829 he showed a rare and total selflessness in his own personal position. Courteault (Courteault 1925, p. 166) gives a favourable sketch of Olinde's character: for him, he was a Jew who 'practiced personally the old Christian virtues of disinterest and charity.' 'He was good, of a candid goodness and of a naive generosity,' a believer in 'the natural goodness of men.' This is of course the view of one who judged Rodrigues by his deeds: some of the Saint-Simonians who opposed him may have seen a different side to his character. Letters from women to *Le Globe* often praise his character. Francisca Prugniaux, for instance, lamenting his departure from the Saint-Simonian church in 1832, wrote: 'It was he [Rodrigues] who has given me much to think about, who has inspired me with ideas which have put me on the road of progress. It is thanks to him that I overcame the prejudices that I have entertained all my life against Jews.'[55] Despite all his manifold activities, Rodrigues was a staunch family man. We have already seen the influence he had on his younger brother Eugène, but he also introduced his cousins,

---

[54]It is worthwhile mentioning that Chevalier's expression is not metaphorical: in (d'Allemagne 1930, p. 364), there is a whole section on the marches organized for the '[*a*]*rmée pacifique des travailleurs.*'

[55]Letter published in *Le Globe* on 27 February 1832. See (Riot-Sarcey 1998, pp. 126–127).

the Pereire brothers, who later became very influential figures in France,[56] to the sophisticated intellectual and banking circles that frequented his household.

Rodrigues is sometimes presented as some sort of a failure because it was clear that he lacked Enfantin's charisma and had thus to yield the leadership of the Saint-Simonians to him. It must be understood, however, that by the time Rodrigues separated from the movement, it had become a totally discredited group, ridiculed by most French intellectuals. It is totally unrealistic to imagine that Rodrigues would have wanted to head such a group: Enfantin's 'victory' was nothing more than the destruction of what could have been a respected force in France's social development. Historians of the French labour movement recognize this: in his comprehensive dictionary of this movement J. Maitron writes: 'More than many other Saint-Simonians, Olinde Rodrigues kept faith with Saint-Simon's desire to improve the fate of the poorest and of the largest number of people.'[57]

We might speculate as to what Rodrigues might have produced had he been able to develop as a full-time mathematician, but as a banker, a financier, and a Saint-Simonian, he prepared the foundations for the just social treatment of workers, for the abolition of slavery, for the rights of women, and for controlling the power of capital, all ideas that informed early socialism. His few mathematical works are of such quality that they alone should ensure that he has a major place in the mathematical pantheon.

Acknowledgements

The authors are grateful to M. Jean Savignac, Mme. Claudine Billoux, and the Secretary of the *Association des anciens élèves de l'École Normale Supérieure* for useful correspondence. Also, the authors thank Dr. Paola Ferruta for details of Rodrigues's marriage certificate and most especially for generously providing us with the Rodrigues portrait in the *Bibliotèque de l'Arsenal.* Professor Jeremy Gray kindly drew our attention to R. P. Locke's book. Mme. Stéphane Toublanc, of the *Cercle de Généalogie Juive* of Paris, gave us very useful information about the Decree of Bayonne, for which we are grateful, as well as for her helpful and patient treatment of our numerous inquiries. The first author is extremely grateful to Professor Paulo Almeida of Lisbon for a very useful discussion on Olinde Rodrigues's works on differential geometry.

M. Louis Bergès, director of the *Archives Départementales de La Gironde* in Bordeaux, has very kindly provided us with a photographic reproduction of the Rodrigues portrait published in 1925 by Courteault in the *Revue Philomatique de Bordeaux et du Sud-Ouest.* We are grateful to Gordon Baldwin, of the Getty Museum in California, and Barbara Tannenbaum, Chief Curator of the Akron Art Museum in Ohio, for information about Felix Nadar, and we are much indebted to Arthur Ollman, Director of the California Museum of Photographic Arts, for putting us in touch with these scholars. Also, we thank Mme. Françoise Reynaud, of the Carnavalet Museum, Paris, for a discussion about the Nadar photograph. Mme. Stéphane Toublanc of the *Cercle de Généalogie Juive* in Paris helped us very much over the years in the use of some of the Paris Archives, and we would like to record with gratitude her kind support and the extremely useful advice that she has

[56]See (Ratcliffe 1971), (Ratcliffe 1972), (Autin 1984).

[57]See (Maitron 1966, p. 332). This biographical article on Rodrigues never mentions his being a mathematician.

given us personally and by correspondence. We are also most grateful to Dr. Jacques Béjot, a member by marriage of the Pereire family, who most generously provided us, through Dr. Paola Ferruta, with large files of genealogical notes on the Pereire-Rodrigues family, prepared by Dr. Patrice Assouad and other Parisian genealogists, some of them members of the Rodrigues-Henriques and Pereire families. We are also grateful to M. Paul Siméon and Dr. Patrice Assouad for discussion on these results.

Bibliography of Olinde Rodrigues's works

These works are listed in chronological order, including those with collaborators. The dates of mathematical works are given in bold.

(1813) *De l'angle de contingence d'une courbe à double courbure.* Paris. ['Question proposée à la thèse de licence, soutenue par M. Rodrigues, le 29 novembre 1813.' See (Hachette 1808). Published in (Rodrigues 1816*b*).]

(1815*a*) *De l'attraction des sphéroïdes: thèse soutenue devant la Faculté de Sciences de Paris par M. Rodrigues, docteur-ès-lettres* [sic], *le 28 juin, 1815*, pp. 27, Paris. [Published in (Rodrigues 1816*f*).]

(1815*b*) Sur quelques propriétés des intégrales doubles et des rayons de courbure des surfaces. *Bulletin Scientifique de la Société Philomatique de Paris*, 34–36. [No volume.]

(1816*a*) Sur le mouvement de rotation des corps libres. *Correspondance sur l'École Impériale Polytechnique*, **3**, no. 1, January 1814, 32–36. [See (Hachette 1808).]

(1816*b*) De l'angle de contingence d'une courbe à double courbure. *Correspondance sur l'École Impériale Polytechnique*, **3**, no. 1, January 1814, 36–37. [See (Hachette 1808).]

(1816*c*) Sur la résistance qu'éprouve un point matériel assujetti à se mouvoir sur une courbe donnée. *Correspondance sur l'École Impériale Polytechnique*, **3**, no. 1, January 1814, 37–39. [See (Hachette 1808).]

(1816*d*) De la manière d'employer le principe de la moindre action, pour obtenir les équations du mouvement rapportées aux variables indépendantes. *Correspondance sur l'École Impériale Polytechnique*, **3**, no. 2, May 1815, 159–162. [See (Hachette 1808).]

(1816*e*) Recherches sur la théorie analytique des lignes et des rayons de courbure des surfaces, et sur la transformation d'une classe d'intégrales doubles, qui ont un rapport direct avec les formules de cette théorie. *Correspondance sur l'École Impériale Polytechnique*, **3**, no. 2, May 1815, 162–182. [See (Hachette 1808).]

(1816*f*) Mémoire sur l'attraction des sphéroïdes. *Correspondance sur l'École Impériale Polytechnique*, **3**, no. 3, January 1816, 361–385. [See (Hachette 1808). This memoir is divided into two parts, Part 1, pp. 361–374, and Part 2, pp. 374–385.]

(1820) and M. Maas. *Théorie de la caisse hypothécaire, ou examen du sort des emprunteurs, des porteurs d'obligations et des actionnaires de cet établissement.* Delaunay, Ponthieu, Pelicier, Paris.

(1825) De l'industrie. Considérations générales sur les banquiers. In H. de Saint-Simon, L. Halévy, and O. Rodrigues, *Opinions littéraires, philosophiques et industrielles*, pp. 161–199. Bossange Père, Paris.

(1829) (ed.) *Doctrine de Saint-Simon. Exposition, première année.* Au Bureau de l'Organisateur, Paris.

(1831*a*) *Religion saint-simonienne. Réunion générale de la famille. Note sur le mariage et le divorce lue au Collège de la religion saint-simonienne... par le Père Rodrigues.* Everat, Paris.

(1831*b*) *Religion Saint-Simonienne. Appel au Bureau du Globe*, Paris.

(1832*a*) *Aux Saint-Simoniens (13 février 1832): Bases de la loi morale, proposées à l'acceptation des femmes.* Everat, Paris.

(1832*b*) *Procès en Police correctionnelle (sous prévention d'escroquerie), etc.*, Paris. [An account of the trial of B. P. Enfantin and B. O. Rodrigues.]

(1832*c*) *Religion saint-simonienne, Olinde Rodrigues à Michel Chevalier, redacteur du Globe.* Lachevardière, Paris.

(1832*d*) *Religion Saint-Simonienne. Procès en la Cour d'assises de la Seine les 27 et 28 août 1832.* Librairie Saint-Simonienne, Paris.

(1832*e*) *Le disciple de Saint-Simon aux Saint-Simoniens et au public.* Everat, Paris.

(1832*f*) (ed.) *Saint-Simon, son premier écrit: Lettre d'un habitant de Genève à ses contemporains, 1820; sa parabole politique, 1819; le Nouveau christianisme, 1825.* Librairie Saint-Simonienne, Paris.

(1838*a*) Sur le nombre de manières de décomposer un polygone en triangles au moyen de diagonales. *Journal de Mathématiques Pures et Appliquées (Journal de Liouville)*, **3**, 547–548.

(1838*b*) Sur le nombre de manières d'effectuer un produit de n facteurs. *Journal de Mathématiques Pures et Appliquées (Journal de Liouville)*, **3**, 549.

(1838*c*) Démonstration élémentaire et purement algébrique du development d'un binome élevé à une puissance négative ou fractionnaire. *Journal de Mathématiques Pures et Appliquées (Journal de Liouville)*, **3**, 550–551.

(1839) Note sur les Inversions, ou dérangements produits dans les permutations. *Journal de Mathématiques Pures et Appliquées (Journal de Liouville)*, **4**, 236–240.

(1840*a*) Des lois géométriques qui régissent les déplacements d'un système solide dans l'espace, et de la variation des coordonnées provenant de ses déplacements considérés indépendamment des causes qui peuvent les produire. *Journal de Mathématiques Pures et Appliquées (Journal de Liouville)*, **5**, 380–440. [English translation in (Baker and Parkin 2003).]

(1840*b*) *De l'organisation des banques à propos du projet de loi sur la Banque de France.* Fournier, Paris.

(1840*c*) *La paix et le travail. Journal Politique et Litéraire.* [Pamphlet n.p.n.d.]

(1841*a*) (ed.) *Poésies sociales des ouvriers, réunies et publiées par Olinde Rodrigues.* Paulin, Paris.

(1841*b*) (ed.) *Une parabole politique.* Capelle, Paris [Oeuvres de Saint-Simon.]

(1841*c*) *Les peuples et les diplomates: La paix ou la guerre, 18 sept. 1840.* Schneider et Langrand, Paris.

(1843*a*) Du développement des fonctions trigonométriques en produits de facteurs binomes. *Journal de Mathématiques Pures et Appliquées (Journal de Liouville)*, **8**, 217–224.

(1843*b*) Note sur l'évaluation des arcs de cercle, en fonction linéaire des sinus ou des tangentes de fractions de ses arcs, décroisant en progression géométrique. *Journal de Mathématiques Pures et Appliquées (Journal de Liouville)*, **8**, 225–234.

(1845) Démonstration d'un théorème connu sur les fractions continues périodiques. *Nouvelles Annales de Mathématiques*, **4**, 109–112.

(1848*a*) *Paroles d'un mort publiées par O. Rodrigues.* Chaix, Paris. [Speeches of C. H. de Saint-Simon.]

(1848*b*) *Projet de constitution populaire pour la République Française, suivi de projets de lois organiques sur la constitution des banques, l'association du capital et du travail, et le mariage; et de développements sur la Bourse et la crise financière; et sur les droits politiques des femmes.* Chaix, Paris.

(1848*c*) *De l'organisation du suffrage universel, proposition d'un nouveau mode électoral.* Saint-Jorre, Paris.

(1848*c*) *Organisation du travail. Association du travail et du capital.* Bière, Paris.

(1848*d*) *Théorie des banques.* Chaix, Paris.

(1848*e*) *Avis aux travailleurs des deux sexes.* Chaix, Paris.

General bibliography

Altmann, S. L. (1986). *Rotations, quaternions, and double groups.* Clarendon Press, Oxford. [Dover Books reprint, NY, 2005]

Altmann, S. L. (1989). Hamilton, Rodrigues, and the quaternion scandal. *Mathematics Magazine*, **62**, 291–308.

Altmann, S. L. (1992). *Icons and symmetries*, Clarendon Press, Oxford.

Altmann, S. L. and P. Herzig (1994). *Point-group theory tables.* Clarendon Press, Oxford.

Autin, J. (1984). *Les frères Pereire: le bonheur d'entreprendre.* Librairie Académique Perrin, Paris.

Baker, J. E. and I. A. Parkin (2003). Fundamentals of screw motion: Seminal papers by Michael Chasles and Olinde Rodrigues. School of Information Technologies, University of Sydney.

Benbassa, E. (1999). *The Jews of France: A history from the antiquity to the present* (transl., M. B. DeBevoise). Princeton University Press.

Bertrand, J. (1902). *Éloges académiques. Nouvelle Série.* Hachette, Paris.

Billoux C. (2001). Personal communication.

Booth, A. J. (1871). *Saint-Simon and Saint-Simonism: A chapter in the history of socialism in France.* Longmans, Green, Reader & Dyer, London.

Carlitz, L. (1970). Sequences and inversions. *Duke Mathematical Journal*, **37**, 138–198.

Cartan, E. (1938). *Leçons sur la théorie des spineurs. I. Les spineurs de l'espace à trois dimensions.* Hermann, Paris.

Cavignac, J. (1987). *Dictionnaire du Judaïsme bordelais aux XVIIIe et XIXe siècles: biographies, généalogies, professions, institutions.* Archives Départementales de la Gironde, Bordeaux.

Cavignac, J. (1991). *Les Israélites bordelais de 1780 à 1850: Autour de l'émancipation.* Publisud, Paris.

Cayley, A. (1845). On certain results relating to quaternions. *The Philosophical Magazine*, **26**, 141–145.

Chambre, H. (1970). Le caractère spirituel du pouvoir chez Saint-Simon. *Cahiers de l'Institut de Science Économique Appliquée*, **4**, 716–729.

Charléty, S. (1931). *Histoire du Saint-Simonisme (1825–1864)*, 2nd ed., Paul Hartmann, Paris. [1st ed., 1896.]

Coignet, C. (1883). Saint-Simon et le Saint-Simonisme. *La Nouvelle Revue: Cinquième Année*, **20**, 125–173.

Courteault, P. (1925). Un bordelais Saint-Simonien. *Revue Philomatique de Bordeaux et du Sud-Ouest*, **28**, 152–166.

d'Allemagne, H.-R. (1930). *Les Saint-Simoniens: 1827–1837.* Librairie Gründ, Paris.

d'Allemagne, H.-R. (1935). *Prosper Enfantin et les grandes entreprises du XIXe siècle.* Librairie Gründ, Paris.

Dupuy, P. and G. Perrot (eds.) (1895). *Le centenaire de l' École Normale 1795–1895* (3 vols). Librairie Hachette, Paris.

École Polytechnique (1895). *Livre du centenaire 1794–1894* (3 vols). Gauthier-Villars, Paris. [Vol. 2, 1894; Vol. 3, 1897.]

Ferruta, P. (2001). Personal communication.

Gerits, A. (1986). Additions and corrections to *Jean Walch Bibliographie du Saint-Simonism.* Gerits & Son, Amsterdam.

Grattan-Guinness, I. (1981). Mathematical physics in France, 1800–1840: Knowledge, activity, and historiography. In *Mathematical perspectives: Essays on mathematics and its historical development* (J. W. Dauben, ed.), pp. 95–138. Academic Press, New York.

Grattan-Guinness, I. (1990). *Convolutions in French mathematics, 1800–1840: From the calculus and mechanics to mathematical analysis and mathematical physics* (3 vols.). Birkhäuser Verlag, Basel.

Gray, J. J. (1980). Olinde Rodrigues' paper of 1840 on transformation groups. *Archive for History of Exact Sciences*, **21**, 375–385.

Hachette, J. N. P. (ed.) (1808). *Correspondance sur l'École Impériale Polytechnique: à l'usage des éleves de cette école* (3 vols.). [Vol. 1, Bernard, Paris, 1808; Vol. 2, Klostermann, Paris, 1813; Vol. 3, *Correspondance sur l'École Royale Polytechnique: a l'usage des éleves de cette école*, Ve Courcier, Paris, 1816; Vol. 1, 2nd ed., Klostermann, Paris, 1813.]

Halévy, L. (1863). *F. Halévy: Sa vie et ses œuvres* (2nd ed.). Heuget et Cie., Paris.

Hamilton, W. R. (1844). On quaternions: or a new system of imaginaries in algebra. *The Philosophical Magazine, 3rd series*, **25**, 489–495.

Hankins, T. L. (1980). *Sir William Rowan Hamilton.* Johns Hopkins University Press, Baltimore.

Jourdain, P. E. B. (ed.) (1908). *Abhandlungen über die Prinzipien der Mechanik von Lagrange, Rodrigues, Jacobi und Gauss.* Engelmann, Leipzig.

Klein, F. (1956). *The icosahedron and the solutions of equations of the fifth degree* (2nd ed.), (transl., G. G. Morrice). Dover Publications, NY.

Klein, F. and A. Sommerfeld (1897). *Über die Theorie des Kreisels*, Vol. 4. Teubner, Leipzig. [Vol. 1, 1897; Vol. 2, 1898; Vol. 3, 1903; Vol. 4, 1910.]

Laptev, B. L. and B. A. Rozenfel'd (1996). Geometry. In *Mathematics of the 19th century: Geometry, analytic function theory* (A. N. Kolmogorov and A. P. Yushkevich, eds.). Birkhäuser Verlag, Basel.

Locke, R. P. (1986). *Music, musicians, and the Saint-Simonians.* The University of Chicago Press.

Maitron, J. (1966). *Dictionnaire biographique du mouvement ouvrier français. Première partie: de la revolution française à la fondation de la Première Internationale*, Vol. 3. Les Editions Ouvrières, Paris.

Manuel, F. E. (1962). *The prophets of Paris.* Harvard University Press, Cambridge, MA.

Marielle, B. C.-P. (1855). *Repertoire de l'École Imperiale Polytechnique, ou renseignements sur les élèves qui ont fait parti de l'institution depuis l'époque de sa création en 1794 jusqu'en 1853 inclusivement.* Mallet-Bachelier, Gendre et succeseur de Bachelier, Paris.

Michaud (1843). *Biographie universelle ancienne et moderne. Nouvelle édition*, Vol. 36. Desplaces, Paris. [1843 is the date of Vol. 1. No date for Vol. 36.]

Muso, P. (1999). *Saint-Simon et le Saint-Simonisme.* Presses Universitaires de France, Paris.

Nahon, G. (1989). The sephardim of France. In *The sephardi heritage: Essays on the history and cultural contribution of the Jews of Spain and Portugal*, Vol. 2 (R. D. Barnett and W. M. Schwab, eds.), pp. 46–74. Gibraltar Books, Grendon, Northants.

O'Donnell, S. (1983). *William Rowan Hamilton: Portrait of a prodigy.* Boole Press, Dublin.

Pereire, E. (1860). *Tables de l'intérêt composé des annuités et des rentes viagères.* Imprimerie Dupont, Paris.

Pinet, G. (1894). L'École Polytechnique et les Saint-Simoniens. *La Revue de Paris*, **3**, 72–96.

Ratcliffe, B. M. (1971). Les Pereire et le saint-simonisme. *Économies et Sociétés*, **5**, 1215–1255.

Ratcliffe, B. M. (1972). Some Jewish problems in the early careers of Emile and Isaac Pereire. *Jewish Social Studies*, **34**, 189–206.

Ratcliffe, B. M. (1995). The Saint-Simonians in the French economy: Toward an understanding of ideas, ideals and action. In *Fra spazio e tempo* (Ilaria Zilli, ed.) (3 vols.). Vol 2, pp. 733–754. Edizioni Scientifiche Italiane, Naples.

Reid, J. (ed). (1902). *Memoirs of Sir Edward Blount K.C.B. &c* (2nd ed.). Longmans, Green, and Co., London.

Riesz, M. (1958). *Clifford numbers and spinors.* Lecture Series, No. 38. The Institute for Fluid Dynamics and Applied Mathematics, University of Maryland.

Riot-Sarcey, M. (1998). *Le réel de l'utopie: Essai sur le politique au XIXe siècle.* Albin Michel, Paris.

Routh, E. J. (1877). *An elementary treatise on the dynamics of a system of rigid bodies* (3rd ed.). Macmillan and Co., London.

Royal Society (1871). *Catalogue of scientific papers (1800–1863): Compiled by The Royal Society of London*, Vol. 5. The Royal Society, London. [Reprinted by Scarecrow Reprint Corporation, Metuchen, NJ, 1968.]

Schoenflies, A. and M. Grübler (1902). Kinematik. In *Encyklopädie der mathematischen Wissenschaften mit einschluss ihrer Anwendungen*, Vol. 4, Part 1, pp. 190–278. Teubner, Leipzig. [Dated 1901–1908.]

Sylvester, J. J. (1850). On the rotation of a rigid body about a fixed point. *The Philosophical Magazine*, **37**, 440–444.

Temple, G. (1960). *Cartesian tensors: An introduction.* Methuen, London.

Walch, J. (1967). *Bibliographie du saint-simonisme: avec trois textes inédits.* Vrin, Paris.

Walch, J. (1970*a*). Qu'est-ce que le saint-simonisme? *Cahiers de l'Institute de Science Économique Appliquée*, **4**, 613–29.

Walch, J. (1970*b*). Les saint-simoniens et les grandes enterprises du XIXe siècle. *Cahiers de l'Institut de Science Économique Appliquée*, **4**, 1757–1797.

Weill, G. (1894). *Saint-Simon and son oeuvre: Un précurseur du socialisme.* Perrin, Paris.

Wilson, E. (1941). *To the Finland station: A study in the writing and acting of history.* Secker & Warburg, London.

Archival sources

Archives de Paris, 18, Boulevard Sérurier, 75019 Paris:
État civil reconstitué

État civil (1860–)
Listes des électeurs
Registres de catholicité
Archives de Bordeaux:
Archives municipales
Archives Départementales, 13–25 rue d'Aviau, 33081 Bordeaux
Bibliothèque, Archives Départementales de la Gironde
Archives de la famille Pereire, Paris
Archives du Baron Ameil, Paris
Bibliothèque de l'Arsenal, Paris, Fonds d'Eichthal; papiers saint-simoniens
Cimetière de Montmartre, Paris
Cimetière du Père-Lachaise, Paris (Registry Office)

CHAPTER 3

Towards a Better Understanding of Olinde Rodrigues and His Circle: Family and Faith in His Life and Career

BARRIE M. RATCLIFFE

Towards a better understanding

When in December 1851 at the age of fifty-six he died at his home in a prosperous residential and business quarter of the capital, Olinde Rodrigues seemed to be a typical representative of the Parisian élite. At his death an inventory was made of his estate, a legal requirement for those who had sufficient possessions to merit the procedure, which only three in ten inhabitants of the city actually did, a sign of the wide disparities in wealth between the better-off and popular classes. It revealed that his apartment had the usual fashionable furnishings of the time: mahogany furniture, bronze mantle clocks, and the cultural symbol that was the grand piano, this one made by a family friend, Camille Pleyel. It further showed that he followed trends among the élite in other ways: he had recently bought land and had a small secondary residence built on it at Herblay in the most elegant outer suburbs to the west of Paris, which had just been linked by rail to the capital and which were attracting more and more wealthy summer residents. The most valuable item in the wealth he left was a share in a stockbroking partnership, which appears to indicate he had followed a career in Parisian finance. His handwritten will, preserved in the same notarial archives as his estate inventory, also suggests that he was a man of his times: he bequeathed as much of his estate as the law allowed to his widow and tried to safeguard the interests of his grown-up children, even requesting that any possibly corrupting books in his library be removed so as to protect the innocence of his still unmarried twenty-year-old youngest daughter. At the same time, in that male-dominated world of the double standard he likely had a mistress, for in a separate codicil to his testament, he asked a friend to make sure that 'Madame D' was given a red wallet with its (unnamed) contents.

Appearances, however, can be a *trompe-l'oeil* for in reality Olinde Rodrigues was anything but typical. These same documents that suggest he resembled others in the Parisian bourgeoisie also show that his fortune was modest when compared even to that of his father, who had left a greater fortune at his death five years previously, and especially when compared with others in that small but growing group of Jewish families in finance and the professions who were profiting from the greater freedoms recently accorded to Jews and the increasing opportunities that resulted from rapid growth in the size and prosperity of the capital in the first half of the nineteenth century. His partnership in stockbroking turns out to be a passive rather than an active one, and his profession on his death certificate was given as man of letters. His investment portfolio did not contain bonds or shares, as those of the wealthy increasingly did (Daumard 1996, pp. 59–145), but only two annuities taken out with the Union Insurance Company run by his friend and fellow Jew Myrtil Maas. Even the apartment he had occupied was modest in size and in rent for it opened onto the busy and noisy entrance to the recently completed Saint-Lazare railway station. Other items among his possessions and clauses in his testament tell us still more unequivocally that he was not as unexceptional as he seemed. He not only possessed a library, which half well-to-do Jewish families and a third of gentile families did at this time, but his was an exceptionally extensive one containing over a thousand books, a collection whose contents ranged from over a hundred volumes on political economy to numerous works in mathematics and science (the inventory, unfortunately, does not record titles). We do learn, though, that among his books was a Bible in Hebrew, suggesting that he was Jewish—as indeed he was—but his library also contained histories of the Papacy and amongst his possessions was a gilt-framed picture of the Holy Family, indicating he might have been—as he also was—an apostate, still a rare status in the Jewish community. His will confirms this because it contained no provisions for the traditional prayers and Jewish burial that seven out of ten wills by members of the Jewish élite still requested at the time, even when in life their authors had ceased to be strictly orthoprax. His library further contained works by Saint-Simon and sets of Saint-Simonian periodicals. Moreover, among his itemized possessions were a portrait of Saint-Simon, a plaster bust of him, and, in a drawer in Rodrigues's desk, his cranium lovingly protected in velvet cloth. A clause in Rodrigues's will is also revealing: it suggests that he had long given modest financial support to Saint-Simon's last housekeeper/mistress, Julie Juliand, for he asked that all her IOUs be burned and that the small pension he had always paid her be continued. Other sources also tell us that amongst his papers were his notes for a study of the life of the master that would later be used by a young assistant to complete the first published biography of Saint-Simon (Hubbard 1857).

All these clues point to a multifaceted career which nevertheless, as shown in Chapter 2, has not been well recognized. This is partly because he left no personal papers or letters: all we have are those scattered in the Pereire family papers, in the Alfred Pereire manuscripts in the *Bibliothèque Nationale*, and in the Saint-Simonian files at the *Bibliothèque de l'Arsenal*. Likewise, he never recorded his relations with Saint-Simon during the last two years of the latter's life, and his modest attempt to reissue Saint-Simon's works (Rodrigues 1832*b*)[1] was overshadowed by the publication by erstwhile followers of Enfantin of the massive forty-seven-volume *Oeuvres de Saint-Simon et d'Enfantin* (*Oeuvres*, 1865–1878), which, incidentally, more faithfully presented Enfantin and his supporters' side in the events leading up to Rodrigues's schism than it did his.

There is a final explanation for the disregard scholars have shown Rodrigues: his later public career in the two decades between his departure from the sect until his death in December 1851 was anticlimactic. He ploughed a lonely furrow, and the seeds he tried to sow did not germinate, at least in his lifetime. While this is true, we might point out in his defence that there were many other 'prophets of Paris' at this time—this apposite term is Frank Manuel's (Manuel 1962)—who believed they

[1]All bibliographical references to 'Rodrigues' indicate papers by Olinde Rodrigues.

had glimpsed the future and attempted to convince usually unheeding contemporaries to follow them. We should also bear in mind that he attempted to promote important causes he had already fought for in Saint-Simonism and which often received short shrift in his time—the cause of workers, women, and international peace, as well as reform of banking. In 1839 he joined the newly established Paris Ethnological Society where his interventions stressed racial equality when many others in the group were busy searching for supposedly scientific explanations for racial inequality and what they believed was European superiority. During the 1840 diplomatic crisis over the 'Near East' that was severe enough to seem to threaten peace (Rodrigues 1840*a*) he attempted unsuccessfully to launch a pacifist journal that would also promote the interests of the popular classes ((Rodrigues 1840*b*) and (*Oeuvres*, VII, pp. 195–196)). As he himself confessed (Rodrigues 1841, p. vi), it was whilst he was trying to garner support for his abortive journal that he came to realise how much attitudes among workers had progressed and how little these resembled the myth of the 'dangerous classes' that threatened the established order then being propagated by most in the French élite (Ratcliffe 1991, pp. 542–574). He therefore edited and published a collection of workers' poetry, much of which was the work of authors who published in the *Ruche populaire* newspaper. Unfortunately, most of the poetry was forgettable, and his collection was duly forgotten. About the same time Rodrigues also began to militate in favour of the creation of pension funds (*caisses de retraite*) for workers (Vinçard 1878, pp. 235–236), and though faltering steps were later to be taken under the Second Empire, his activities did not yield immediate results. Finally, in 1848, in those heady days following the February Revolution, when some hoped for a new world, Rodrigues acquired renewed energy. He called a meeting of ex-Saint-Simonians at his apartment (though, significantly, he did not invite Enfantin) in the hope of rallying them to action with other socialists. After stormy exchanges, they failed to agree on what they could do together, and the meeting broke up in disarray (Vinçard 1878, pp. 257–259). He even improvised speeches at the Temple of Mammon, in the wrong-headed hope of winning over to the cause of radical reform those on the floor of the Bourse, and he addressed the Chartists in London as discussed in Chapter 2. He even tried to gain election to the Assembly and, more interestingly, because it showed again that he stayed faithful to the forward-looking views of his Saint-Simonian years, he advocated that universal suffrage, reestablished for men in March 1848, be extended to women (Rodrigues 1848*a*, 1848*b*). In his time only a handful of avant-garde militants listened to his call.

His role in Saint-Simonism was, admittedly, partly elusive, but the experience marked him. As Eugène Lerminier, who had also belonged to the sect, said in his old age, '[l]et's admit that it was the only beautiful moment in our life' (see (Ratcliffe 1971, p. 1226)). At the same time, Rodrigues may well have marked Saint-Simonism in a more significant manner than he is given credit for. First, as is generally agreed, he provided moral and probably modest financial support to Saint-Simon in the last months of his life, a time when the philosopher, who had always been better able to think than to write, published his important *Nouveau christianisme* (Saint-Simon 1825) and he, Olinde Rodrigues, and another Jewish friend, Léon Halévy, published *Opinions littéraires, philosophiques et industrielles* (Saint-Simon *et al.* 1825). Though we cannot establish with certainty the contributions made by Rodrigues and Halévy and though the latter subsequently proved he had a facile pen whereas the former did not, it does seem that, at the very least, Rodrigues wrote the foreword to the *Nouveau christianisme* and two chapters in *Opinions littéraires philosophiques*, one of which was on banking and the role it might play in the future (Rodrigues 1825, pp. 161–199). Second, again as discussed in Chapter 2, he was the catalyst for the movement that, after the master's death in 1825, took up—and developed—Saint-Simon's ideas. He also played an active role in recruiting adepts and discussing theories during the silent gestation period before the movement acquired more momentum and a public profile from 1829 onwards. Thirdly, though it is always difficult to establish paternity for ideas, and especially in this case, because the issues he stressed were to be found in Saint-Simon and would also be emphasized by other Saint-Simonians, he was instrumental in stressing the themes of international harmony, the role of banking, and equality for women and for the popular classes.

A third facet of his career also makes Rodrigues worthy of notice: his contributions to thinking about business and especially banking and, to a lesser extent, insurance. Scholars have always made much of the importance of Saint-Simonian ideas on credit and banking. They are wrong to do so because these ideas were less original than they suppose, and apart from the breadth of vision in their emphasis on an integrated and highly specialized credit system serving the more than economic ends they ascribed to it, Saint-Simonians made no contribution to banking theory, and only one of the techniques they put forward has any claim to originality: their proposal for the issue of interest-bearing banknotes as a means of extending the credit network (Ratcliffe 1975*a*, pp. 23–47). We must accept, though, that, in common with others of their contemporaries, they put great faith in banking as an instrument of change. Even here, however, it is extremely difficult to determine the precise role that Rodrigues himself played in developing Saint-Simonian thinking because (except for (Rodrigues 1825)) the most important texts that were published were written by others and particularly by Enfantin and the Pereires. It is clear, however, that Saint-Simonian ideas emerged out of exchanges within the group and with others in Parisian finance and that Rodrigues played a leading role in these discussions. Moreover, he stayed faithful to these ideas, as did his cousins, the Pereire brothers, for he again took up criticisms of the Bank of France in the 1840s (Rodrigues 1840*c*, 1848*c*). In both of these texts he repeated criticisms made earlier of the Bank of France for what he and like thinkers always regarded as its oligarchic structure, timorous policies, and failure to assume a more dynamic and truly national role in the economy. Rodrigues also made more direct contributions to the financial system. In 1820 he and his fellow mathematician Myrtil Maas made a detailed critique of the functioning of a recently formed and pioneering banking institution, the *Caisse hypothécaire* made a detailed critique of the functioning of a set up to offer mortgage loans to farmers (though in practice it made fewer of these than it did loans for urban development). They criticized its organization and used their mathematical training to estimate rates of return for its shareholders and bondholders (Rodrigues and Mass 1820, pp. 31–39). Rodrigues was to be rewarded by being named deputy director of the institution and then its director.

There is a fourth and final manner in which Rodrigues is worthy of study: for what the sea-changes, political, socio-economic, and cultural, affecting the Jewish minority in France in the late eighteenth and early nineteenth centuries can tell us about his life and for what his family and his Jewishness can tell us about attitudes and comportments among the Jewish financial and professional élite in the capital. Olinde Rodrigues was born into the closely knit Jewish business élite in the port city of Bordeaux where intermarriage and family transmission of occupations helped ensure both prosperity and continuity. The economic crises faced by the Atlantic trade of the port, the Rodrigues family's decision to move to Paris, and, above all, the emancipation of the Jews that was instituted only five years before his birth and the intellectual ferment and debates about acculturation that took place within the rapidly growing educated Jewish élite in the capital all make this a pregnant moment in the history of French Jews and these changes central to an understanding of Olinde Rodrigues. The problematic that this chapter sets for itself, indeed, is to gain a better understanding of how his family adjusted to change and what role family and faith played in his life. In so doing, however, we are not suggesting that Rodrigues was merely the product of his family, upbringing, and cultural and ethnic milieu, for this would be to overlook individual springs of thought and action. It is, though, an approach that promises to illuminate. In recent scholarship greater stress has been put on the resilience and therefore continuing importance of family ties. We also know how significant community ties can be because scholars have recently put increasing emphasis on the importance in any society or group of what has come to be termed the social capital that community constitutes (the concept of social capital was first put forward by James S. Coleman (Coleman 1988, 1990)). What happened to family and community in post-emancipation Paris, and how did stability and change affect Olinde Rodrigues?

Answers to these questions will be forthcoming only if sources are available and can be properly interrogated. Unfortunately, the sources available for a study of Olinde Rodrigues, his family, and his circle are especially lacunar for no personal papers or correspondence have come down to us. We are therefore obliged to consult the less obvious, and in many ways less satisfactory, traces they left in their wake. Besides a reading of published materials, our research strategy chiefly has consisted of piecing together clues scattered in four very different types of documentary sources. To avoid overburdening the text with references to this patchwork of sources, only published documents and quotations from unpublished documents will be cited. Instead, the principal elements in the database will be described here. The first is family papers and correspondence which regrettably make some but still too infrequent references to Rodrigues and his family.[2] The second consists of documents from or about the Jewish community in Paris and especially the Consistory, which since 1809 has been responsible for community affairs.[3] One Consistory document, the 1809 census of the Jewish population of

[2]The Pereire family papers, which are in private hands, contain some correspondence and manuscripts from the Rodrigues family that we used in our earlier studies of the Pereire brothers. The papers of the Saint-Simonians (the *Fonds Enfantin, Fonds d'Eichthal*, and *Fonds Lambert*) are to be found at the *Bibliothèque de l'Arsenal*, while other papers (which probably all come from Olinde Rodrigues's library) have been deposited as the *Fonds Alfred Pereire* in the *Département des manuscrits* of the *Bibliothèque Nationale*. Some boxes of d'Eichthal's papers are to be found at the *Bibliothèque Thiers* (including brief biographical notes on Olinde Rodrigues in *carton IV*). As we shall explain, useful as they are, none of these is as valuable as might be expected.

[3]Two series are available for the study of the Paris Consistory. One is the F^{19} series from the Ministry of Religion (*Ministère des cultes*). Boxes F^{19} 11004, 11010, 11028, and 11042 were especially useful. The other is the holdings of the *Archives du Consistoire* on the rue Saint-Georges. Because of the German occupation of Paris in World War II and because of the depredations caused by rodents and humidity in the cellars of the Rothschild château where parts of the archives were

the capital made at the behest of the government, which used to be in its archives has long been missing but has now mysteriously turned up in the holdings of the Jewish Theological Seminary in New York. We have been able to analyse this register which proved particularly useful for establishing spatial and occupational distribution, family size, and date at which immigrant heads of households arrived in Paris.[4] The third consists of serial sources containing data collected for other reasons which, when used in conjunction, yield information on the addresses, types of residence, occupations, and wealth of the Rodrigueses and their circle.[5] The fourth is the most promising: registration records of vital events in a person's life, that is, birth, marriage, and death.[6] Apart from furnishing obvious personal information, such documents, especially when juxtaposed with others at the time, allow scholars to read them for ends other than those for which they were drawn up and thereby uncover regularities and patterns. We have already intimated what a close reading of estate inventories and wills might yield about consumption patterns, family ties, and individual lives. The same is true for birth certificates, which not only indicate the address and occupations of parents and the two witnesses, but in the case of forenames chosen for the new-born, something about attitudes. Also useful are marriage contracts, signed by parties before a wedding involving the better-off, which are conserved in the notarial records deposited at the *Archives Nationales*. They indicate, for instance, the marriage portions brought to the union by bride and groom, and these suggest not just the wealth of future partners but the nature of the family alliance that marriages, especially but not exclusively among

---

hidden at that time, holdings are lacunar. We used the AA 2 and 3, GG 1 and 2, and the CC 33 registers and correspondence. Regrettably, registers of circumcisions and burials are completely missing for our period, while, in contrast to the complete and detailed Catholic parish registers, surviving marriage registers are both laconic and lacunar.

[4]The register is now held among the *pinkassim* of past Jewish communities in the archives of the Jewish Theological Seminary. There is good reason to believe that this census offers a more accurate image of the élite than it does of the popular classes. It is only a freeze-frame of a moving reality because, especially in the case of the mass of the Jewish population who were peddlers, migrants moved in and out of the capital, and the presence of many of them likely was not recorded in the register or, if registered, only temporary.

[5]At the *Archives de Paris* the cadastral records (DQ^{18} *sommier foncier* and D^1P^4 *calepins de révision du cadastre*) were consulted for the buildings the Rodrigueses inhabited at different times in the capital. These indicate the nature and ownership of buildings, as well as the number and value of rental units. We did not find any leases for the Rodrigueses probably because leases are only unevenly indicated in the registers. We know the addresses where they lived at different times from other sources. Also at the *Archives de Paris* are electoral lists for the period (VD^4 and D^2M^2 series.) These indicate who had the suffrage in a system where qualification for the vote depended on taxes paid, and they further give the addresses, age, occupation, and the taxes paid by each qualified elector. Both Isaac and Olinde Rodrigues are absent from all available lists of electors in the first half of the nineteenth century. Olinde Rodrigues is only included in 1848 when universal manhood suffrage was instituted. Their absence means, of course, that neither paid enough taxes to qualify to be electors.

[6]The fires set by Communards in May 1871 consumed all civil registration records for Paris before 1860. While the loss of all these is irreparable, the *Archives de Paris* has two incomplete sets of reconstituted records (*État civil reconstitué*) available on microfilm (5mi 1), which were used to trace at least some of the vital events in the lives of the Rodrigueses and other families in their circle, as well as *registres de catholicité* and Catholic Church registers (*Archevêché*) which were used to trace converts to Catholicism. The extremely rich Minutier central at the *Archives Nationales* contains the great bulk of notarial acts that were drawn up at this and other periods. For the Rodrigues and Pereire notarised documents, the records of the notaries Bertrand and Fould proved especially useful.

the better-off, constituted. For their part, marriage certificates, the most detailed of all civil registration documents in France, yield vital information about spouses and parents and the four witnesses that the Civil Code obliged each couple to provide (and these could not be parents or females). However, they indicate much more and allow the historian to learn about wider social processes. We have tried to do as much as possible along just these lines to better understand the role played by family and faith in Olinde Rodrigues's life and in those of his immediate circle.

Family

To establish something of the role of family in his life is to do more than gain a better understanding of Olinde Rodrigues: it is also to raise wider issues because a study of family sociabilities, solidarities, and conflicts is a litmus test for measuring the extent of stability and change among the Jewish élite in Paris during the half century following emancipation. It is not surprising that the Abbé Grégoire, one of the principal promoters of the emancipation decree of 1791, who was highly critical of many characteristics of Jewish communities at the time, was also a great admirer of their family life, one of their sources of strength.

Nor is it surprising that in the upper echelons of the Sephardic community in later eighteenth-century Bordeaux family was not just the locus for the socialization and education of girls and partly boys, but it was also the centre of social and much of cultural life. Occupations were transmitted through the family because boys early entered their fathers' businesses or did apprenticeships in the same trade and because, where possible, businesses were handed down from fathers to sons. Marriage patterns also reinforced the oligarchical character and success of the business élite because over four out of five weddings in the Bordelais community at the end of the century involved spouses from within the same occupational milieu. This pattern of endogamy also extended to residence—before marriage nine out of ten spouses lived in the same quarter, the same street, or even the same house—and to subgroups because Sephardim did not marry into the Ashkenazim or other minorities in the Jewish community, though we should note that this was not because of ethnic differences but because other groups, which were more recent arrivals in the city, were usually in other occupational and social milieus (Cavignac 1991, pp. 275–283). Among the élite the only exceptions to this residential propinquity of spouses were alliances between families in business in Bordeaux and the nearby Bayonne, which did not break the endogamous tradition. Such family ties and intermarriage help explain the success of the Jewish business élite and their importance in strategic sectors of the Bordelais economy: among merchant brokers their domination was complete; among marine insurers, commission agents, bankers, and shipowners they had a significant share of business (Cavignac 1991, p. 102).

As were those of others, the family life of two generations of the Rodrigues family, that of Isaac Rodrigues and his wife Sara Sophie Lopes Fonseca and their seven children and that of Olinde Rodrigues and his six siblings and their children, was affected by two radical changes. The one promised to impact all families in the French Jewish community: emancipation transformed the environment in which families functioned. With the altered legal system and the loosening of community ties, acculturation became easier. Attending gentile educational institutions now became possible, and the better-off could avail themselves of the opportunity. A wider range of occupations was now also open to Jews. Constraints on choice of marriage partner were also lifted because under the new legal code civil weddings not only became possible but obligatory, while religious ceremonies were rendered merely optional. Though at the critical life junctures of birth, marriage, and death, all of which required civil registration, nine out of ten in the general and Jewish population were long to continue to have recourse to religious ceremonies. Henceforth the way was open for couples to opt only for legal registration, and mixed marriages between Jewish and gentile partners became a possibility. The second major change also affected the Rodrigues family, as it did others who migrated to the French capital: the different conditions that prevailed in Paris in the decades following emancipation. We should not expect, however, that these circumstances would inevitably transform family relations in the short run. This is not just because there is reason to believe that, under the exceptionally liberal circumstances the Sephardic élite in Bordeaux enjoyed and under the influence of enlightenment thinking, some gradual changes had already occurred in family life, but also because scholars now realize better than they once did that, though observers of modernization, then as now, frequently lamented what they thought was the ominous erosion of family values and structures, especially among migrants and the popular classes, families everywhere have proved surprisingly resilient whatever the outside challenges they have had to face. In any case, as we shall see, the Rodrigues family in Paris faced conflicting rather than one-way pressures.

The fact remains, of course, that the city to which the Rodrigueses moved in 1796 was different in size and in economic structure from the city they left. The Paris Jewish community, though when they arrived was still comparable in size to that of Bordeaux, would differ in the rapidity of its subsequent increase and in its composition, for it attracted immigrants from other communities, and especially Ashkenazim from eastern France. Three quarters of the nearly three thousand Bordelais Jews in the late eighteenth century were Sephardim, whereas, a decade after the Rodrigueses' arrival in Paris, Jews there from Bordeaux or Bayonne, where there was another interrelated Sephardic community, still numbered only ninety-two and made up just six per cent of those in the community who had not been born in the capital (Kahn 1885, pp. 65–78). Even the Jewish business élite which the family joined would be made up predominantly of immigrants from Alsace-Lorraine and, to a lesser extent, from Central Europe. In contrast to the well-established Sephardim in Bordeaux, the community they came to was also a recent one: the 1809 consistory census indicates that among heads of household not born in the capital—they made up the vast majority of Jewish residents—only one in twenty had come to Paris before 1770, while three in ten had come in the two decades immediately preceding the Revolution, a further third in the 1790s, and another third in the first years of the nineteenth century. This dynamism would persist: not only did the number of Jews in Paris increase much faster than did the population as a whole, which itself doubled in the first half of the nineteenth century (Piette and Ratcliffe 1993), but the numbers and wealth of the Jewish élite grew even faster. Whereas the latter made up just over twelve per cent of the community in 1809, they represented seventeen per cent in the 1840s (Piette 1983, pp. 58–59). At the same time and thanks to freer conditions and changing aspirations, the occupational makeup of this élite was also undergoing change. Between 1809, when, because the government ordered all Jewish communities to carry out a detailed census of their members, we have a first indication of the size and structure of the Jewish population, and mid-century, when a further census was made, the number of Parisian Jews in the liberal professions increased three times (Piette 1983, p. 63; Cohen 1980, pp. 350–351 and 372). Not only were the size and composition of the Jewish élite changing, but so was its spatial distribution across the city. Whereas the bulk of the Jewish population, and particularly the poor, settled in the fourth *arrondissement* (the numbering of Paris districts was modified in 1860; we adopt here present-day numbers), the élite settled in the wealthy but less densely populated quarters in the northwest of the city, which were also being rapidly colonized by Parisian business and bourgeoisie.

They had left Bordeaux partly because the prosperity of the port that had been so remarkable in the second half of the eighteenth century had already been compromised by the Revolution and wars. The city's trade would be further hurt in the following decade and Isaac Rodrigues's closest friend, Isaac Pereire, who married his wife's sister, and who was in the business of marine insurance, would be forced to declare bankruptcy in the difficult conditions imposed by the British naval blockade in 1805 (Ratcliffe 1972, pp. 189–191). Continuing economic troubles explain why Rodrigues was not alone in coming to Paris. Other family members also did so: Isaac Patto came to the capital in 1794, while Benjamin Rodrigues, from another branch of the family, arrived in 1801, and they then set up a banking house together. In the early nineteenth century other Rodrigues also arrived to become stockbrokers at the Paris Bourse. After Isaac arrived in the capital, he set up residence, neither in the St. André-des-Arts quarter in the present-day sixth *arrondissement* on the Left Bank, where Sephardim in eighteenth-century Paris had usually established themselves (Neher-Bernheim 1983) (and where the 1809 census indicates nearly half all those born in Bayonne or Bordeaux still lived), nor in the crowded city centre where the mass of the impoverished Jewish population lived, but in the emerging and prosperous area that is now in the ninth *arrondissement* where many in the gentile and Jewish business class were already establishing themselves and where in the previous year Berr Léon Fould, who was only four years his elder, had set up his banking house. It may well be, indeed, that Fould called Rodrigues to Paris, for the latter rented an apartment on an adjoining street to where Fould lived and had his bank. He was to be associated with this banking house for much of his career.

In such changed and changing circumstances acculturation accelerated. The most obvious change, but not the most significant, was the adjustment of the family name and the forenames given to their children. Unlike some others, they did not have to adopt surnames for the first time nor, as did Isaac's friend Elyahu Halphon Lévy, who became Élie Halévy, Gallicize the one they had. Nor did they choose to change the family name, as another family friend in Berlin, Abraham Mendelssohn, who famously and unsuccessfully attempted in 1829 to persuade his composer son, Felix, to abandon his Jewish-sounding last name in favour of Bartholdy, the name of a protestant in-law. What they did do, though, was to quickly give up in everyday dealings and as authors the second part of the full family name, which was Rodrigues-Henriques. Much more revealing of acculturation, however, are the forenames parents gave their children. The convention in Jewish families had been to confer the names of grandparents or names taken from the Bible. The tradition among the Jewish business élite in eighteenth-century Bordeaux had been to give their children two forenames: a gentile forename to facilitate business contacts with the Christian world in the port and in the Caribbean colonies with which Bordeaux had close ties and a Jewish forename. Isaac Rodrigues, whose given names were Jean Isaac, and his wife, Sara Sophie, observed this convention by giving all their children these two kinds of first names. However, as did others in the Paris Jewish élite at this and later times, the children would always use their Christian rather than their Jewish forenames. Benjamin Olinde, then, was always known as Olinde. The next generation went further: of the twenty-seven children born to Olinde and his siblings only two had traditional Jewish names, and these were Jacob Henry and Isaac Émile, sons of Émile Pereire and Henriette Rodrigues, whose names recalled the paternal grandfather, who had died young and in penury, and great grandfather, whose claim to be the inventor of the first method for teaching deaf-mutes to communicate had been usurped and whose memory Émile Pereire long attempted to rescue. The choice of non-Jewish forenames that Isaac Rodrigues and his wife made for their offspring, as already intimated in Chapter 2, may well be especially revealing because at least four of these suggest just how immersed the parents already were in the wider cultural milieu of their time. As we shall see, acculturation was only one explanation for this abandonment of Jewish forenames.

There is a second sign of acculturation: family size. Jewish families, even among the business élites of Bayonne and Bordeaux, still had larger families than did gentiles in comparable situations ((Oukhemanou 1998) has analysed family size in the Sephardic community in Bayonne in 1808). Such large families may have resulted in part, of course, from choice, such as an attempt to secure a male heir or to replace a child lost in infancy, but they were mainly the result of the absence of family limitation. As the Rodrigues family tree indicates (see Chapter 2), Olinde Rodrigues's paternal great-grandparents had nine children, his paternal grandparents seven, and he himself had seven siblings. However, out of all the latter who married, only one, Sophronie, the eldest daughter, had a large family: seven, including twins from her last pregnancy. Part of acculturation, then, was that, as did others, the Rodrigueses aligned themselves with the family limitation long practiced among gentiles.

There is a third sign of change, but one which is more difficult to measure: the place of wives and daughters in the family. Symptomatic of the subordinate role of Jewish women in eighteenth-century Bordeaux and Bayonne is their capacity to sign their names on official documents compared with that of men. When Jewish families were enumerated in Bayonne in 1808, seventeen out of twenty men proved able to sign their names but only one in six women did so (Oukhemanou 1998). A bride's signature in wedding contracts and registers suggest, of course, that among Jewish élites the percentages of women who could sign were higher but that they were still lower than in comparable social strata in French society as a whole. Though we know very little about women in the Rodrigues families in Bordeaux or even about Sara Sophie Lopes Fonseca, who married Isaac Rodrigues in 1794, we do have some revealing details about her sister, Henriette, younger by a year, who was the mother of the famous financiers Émile and Isaac Pereire. In the notarial records in the *Archives Nationales* is a document called an *acte de notoriété* in which witnesses swore to the date and place of birth of Henriette Pereire because she had no birth certificate, and she did not have one, as the witnesses explained, because in Jewish families the births of boys were recorded but those of girls were only rarely reported, as the register of Sephardic Jews in Bordeaux from 1738 to

1792 indicates.[7] Her correspondence with Émile Pereire, preserved in the family archives, also reveals that, though she came from a prosperous merchant family, she only wrote with difficulty and with grammatical errors. It further shows that, as did so many Jewish mothers at this time, she remained pious and orthoprax and down to her death in 1827 worried about the threats posed by her sons' contacts with the Rodrigues family in Paris because the family had ceased to be observant. When in 1824 Émile Pereire married the second of their daughters, they married religiously, the only marriage in the family that took place in a synagogue. The contrast between Henriette Lopes Fonseca and Olinde's sisters is striking. Some of them were to play a role outside the domestic sphere in the Saint-Simonian movement, where among other things, they defended women's rights, and where their few surviving letters reveal not just their literacy but their wide culture. Their choice of marriage partner also suggests that they played an independent role in the decision as to whom to marry. That two daughters married Christians may indicate the freedom they enjoyed because the great majority of mixed marriages at this time involved Jewish men who had greater contacts with gentile society rather than Jewish women.

There was a fourth and more obvious sign of change: Isaac Rodrigues's children received a different kind of education from that which he and his wife had had themselves. On his own admission, Isaac Rodrigues had begun working as a bookkeeper for the Bordelais merchant André Aquart at the age of ten, which meant that by the time he left for Paris in 1796, he already had fifteen years' experience in accounting. His daughters, however, were sent to private girls' schools in Paris, the older ones possibly to the *pensionnat* (boarding school) run by Henriette Mendelssohn, the youngest daughter of Moses. His sons may have been sent to a *heder* (a Jewish primary school) as were the two sons of Élie Halévy, Fromental, and Léon, before they sent the one to the Conservatory of Music and the younger to the *Lycée Napoléon* (Halévy 1863, pp. 7–8). We know (Chapter 2) that both Olinde and his brother, Eugène Rodrigues, who was seven years younger, went to Paris *lycées*. Though attendance at *lycées* and institutions of higher learning would become commonplace for sons of the Jewish élite, we should note that the decision by Isaac Rodrigues to send his sons to schools in the public system was a novel one and that it reflected the enthusiasm in his circle for the educational and career possibilities that the Revolution and emancipation had opened up. Evidence of this is the case of Élie Halévy, who sent his sons to prestigious schools, not because it was commonplace—it was not—and not because he could well afford to do so: he did so only at considerable expense and despite his own modest means. There are a number of other instances of expressions of enthusiasm for education. The young banker Louis d'Eichthal, who came to Paris from Munich in 1797 to serve an apprenticeship at the Jewish banking house of Worms de Romilly, was excited by the turmoil of ideas he found in the capital, attended scientific lectures, and was struck by the status enjoyed by scientists in France. Henceforth, he would nurture the ambition that at least one of his sons would follow a scientific career and, if possible, gain admission to the prestigious *École Polytechnique* engineering school (Ratcliffe 1977, p. 115). Olinde Rodrigues, then, was one of only a handful of Jews to enroll in public schools. In the year he began studying at the *lycée*, the Paris

[7]We owe this to Stéphane Toublanc, who kindly shared her unpublished research findings with us.

Consistory informed the Ministry of Religion that there were already six Jews in *lycées* in the capital. The 1809 census indicates that at that moment there were four Jewish students studying at the *École Polytechnique*, two in medical school and two at the *Conservatoire*. By 1815, indeed, a dozen Jewish students had already attended the *École Polytechnique* established only in 1794 (Piette 1983, p. 75).

Still more telling about family life and sociabilities and the Rodrigueses' status in the capital are the marriage patterns of Olinde and his sisters between 1815, when the eldest daughter married, and 1838, when the last daughter did. If we are to understand these patterns, we should bear in mind that the pool from which partners could be drawn was limited. Mutual affection counted, but weddings in their social group were also alliances to which the contracting parties had to bring their financial contribution and, in the case of the groom, prospects. As Adolphe d'Eichthal, elder son of Louis, who was still only modestly wealthy, even though he was destined for an exceptional career in banking and to be the first Jew to be named a regent of the Bank of France, put it in a letter to his brother Gustave in 1827:

> You may call me romantic or what you will but... purity... virginal simplicity will be an essential quality in the one I marry. As for wealth, I shall not marry a woman who has nothing, though I shall not insist on a massive fortune ['gros lot' in the original]: a man must know how to be happy in this world....[8]

The choice of life partner that Olinde Rodrigues and his sisters could make had these and still further constraints. One of these is that their family was Jewish and yet, as we shall need to explain, had ceased to be observant and progressively marginalized itself within the community. Another was that their father had only modest wealth and, at times of economic difficulty, had even less available to endow his marriageable offspring. Marriage contracts in notarial records and, in the case of three marriages, their absence, allow us to weigh the importance of this impediment. Scholars have already made some attempt to gauge the average funds that both spouses brought to a marriage both among the Parisian Jewish élite and in the bourgeoisie as a whole at this time, which means that we have points of reference against which to determine the relative level of fortune of Rodrigues's wedding settlements. In two-thirds of the marriage contracts in the Jewish bourgeoisie the joint contribution of the marriage portions of both bride and groom was over ten thousand francs (and in two-fifths of them it was more than twenty thousand) (Piette 1983, pp. 67–70). In only one Rodrigues contract, however, did this contribution exceed ten thousand francs: when Noëmi Amélie married the mathematician Serve-Dieu Abélard Lévy in 1838, the combined total reached just over twenty-nine thousand francs, but this was only because Lévy was a forty-four-year-old widower in mid-career, rather than standing on the threshold of his working life, as did most grooms. Other Rodrigues marriage settlements, however, were only either very modest or nonexistent. In part, this was because future husbands came without fortunes. David Andrade, who married the eldest daughter in 1815, was the son of a rabbi, and rabbis were notoriously ill paid at this time. Émile Pereire, who would become extraordinarily wealthy as early as the beginning of the 1840s, brought only three thousand francs in personal savings when he married the

[8] *Archives Nationales.* F^{19} 11004.

third daughter in 1824. This was because his widowed mother had struggled to make ends meet after his father had died shortly after declaring bankruptcy and was not able to endow the union and because he had only begun his career in Paris the previous year. When, three years later, the second daughter married Romain Regnault, who was destined for a successful career as a general practitioner in a wealthy quarter of the city, he was still finishing his medical studies and, perhaps because his Catholic parents, who lived near Chartres, cut him off (the marriage certificate shows they failed to attend the ceremony), brought no monetary contribution to the marriage at all, an extremely rare occurrence in the Rodrigueses' social group.

Irrespective of these limits to the marriage market that was available to them, choice of life partners nevertheless reveals the very different kind of society in which the Rodrigues family now moved. Only two of the spouses came from the traditional Sephardic business élite pool from which partners had always been drawn. One was David Andrade from Bayonne, whose father had long known Isaac Rodrigues and had been made chief rabbi in Bordeaux. Andrade would have a modest career in Parisian finance. The other was Émile Pereire, a cousin from Bordeaux, whose family had so intermarried with his bride's family over the years that his full family name had become Rodrigues Pereire. In the four other marriages, however, the centuries-old tradition of endogamy was set aside. Three chosen partners were in the liberal professions. As we have seen, one of these was a physician and one was an academic. The third, Henri Baud, who married Rebecca Mélanie in February 1832, was a lawyer at the Paris bar. The fourth life partner was Charles Sarchi, one of three Jewish brothers who had come to Paris from the Austro-Hungarian Empire in 1814 and who held a law doctorate but was to have a business career. The first and most radical break with the past, though, was made by Olinde Rodrigues when in October 1817 he married Victorine Denise Martin, daughter of a Catholic schoolteacher in Paris who had originally come from the Vosges. All four parents of the pair consented to and attended the ceremony, but no marriage contract was drawn up, probably because neither bride and groom nor parents had the financial resources to endow the union. It was an exceptional choice of partner, and we shall attempt to offer an explanation for it.

If our sources permit us to glimpse changes in Rodrigues family life, they also hint at three constants which indicate the resilience and continuing importance of family in the lives of its members. One of these is the occupations taken up by men. Henri Germain, the founder of the Crédit Lyonnais bank, once averred: 'business is not figures, it's men' (Bigo 1947, p. 260). If he had been referring only to the Jewish business élite at this and other times, he might well have said that business was families. Male offspring were early initiated into the family trade. Businesses were handed down from generation to generation and their viability reinforced through marriage alliances with other Jewish families in the same or related trades. Thus in Bordeaux the Rodrigues family had long been bankers or merchants and had followed the familiar pattern of intermarriages in their community. The family tree we have recreated shows both a continuity in occupations—though we should remember that the term 'merchant' covers a range of conditions and that another term '*négociant*' describes businessmen of a higher status—and intermarriage. Similarly, when Isaac Rodrigues married in November 1794, his father described himself as a merchant. His bride came from a Bayonne merchant family living next door to the

Rodrigueses in a house her groom's parents owned, and her father Mardochée Lopes Fonseca, always known as Fonseca *neveu*, was a commission agent and insurance underwriter. The marriage contract (a copy of which is in the notarial archives in Paris), we might add, shows the strains placed on the two families' businesses by the vicissitudes of war and political upheaval: while his parents promised to give the pair twelve thousand livres, they were unable to do so immediately, and her parents, who were able to give them ten thousand livres, paid them in the collapsing Republican paper money, the *assignats* that other sources indicate were then worth only a quarter of their face value.

Not all branches of the family were equal, however, for, as were his brother David and a cousin, Isaac Rodrigues held only a subordinate position as a bookkeeper. When he came to Paris, he continued to work as an accountant, taught what he knew to others in high finance, and published a well-received manual, the first to give instruction in the use of triple-column accounts (I. Rodrigues 1810). He was also employed by Berr Léon Fould and, most importantly, was named chief accountant of the metallurgical firm, the Fonderies du Vaucluse, in which Fould had invested heavily and already made important losses. Rodrigues, then, was there to watch over Fould's interests, which he did for nearly a quarter of a century until 1831 when the company suspended payments (for accounts of Fould's involvement in this enterprise, see (Barbier 1991, pp. 51–72), (Locci 1988, pp. 84–97), and (Lévy-Leboyer 1964, p. 431). He also acted as a kerb-broker at the Paris Bourse, an occupation in which his nephew Émile Pereire, in partnership with his uncle, began his career in Paris in 1823. The scions of other branches of the Rodrigues family who came to the capital similarly took up occupations that followed in the family tradition. The brothers Edouard and Henri Aron became stockbrokers, as did the latter's son (Ratcliffe 1972, p. 194). Hippolyte Rodrigues, from another branch, son and grandson of Bordelais stockbrokers, who really wanted to be—and later became—a writer, also followed family tradition by working at the Bourse (H. Rodrigues 1893). Olinde Rodrigues would have preferred an academic career, but circumstances forced him, too, to take up a career in finance, acting first as deputy director, then director, of the *Caisse Hypothécaire*, a major new banking institution in the capital, and, following his break with Saint-Simonism, working more modestly as a kerb-broker at the Bourse. Family traditions and connections, then, played a part in determining his career, as they did for others in the wider Rodrigues family and Jewish community.

There was a second constant: family coresidence. This had long been a characteristic of family ties, and a feature of marriage settlements in the Jewish élite, in Bordeaux as in Paris, was that parents offered newlyweds food and lodging with them. When Isaac Rodrigues married, for instance, his parents offered the couple lodging in the house next door, which they owned and which was occupied by his bride's parents. The tradition was maintained in the capital. When Henriette Rodrigues married Émile Pereire, her parents offered them, as part of her dowry, food and lodging in their modestly sized apartment. Property taxation and civil registration archives also allow us to establish that, except for brief moments in late 1831 to early 1832 and in the early 1840s, Olinde Rodrigues always shared addresses with his parents. The same cadastral records which we have used to establish this coresidence also show that, in contrast to investment patterns for the Paris bourgeoisie as a whole (analysed in (Daumard 1973)), neither Olinde nor his father ever purchased property in the capital, always rented, and constantly moved, as did the daughters and their families. This does not mean, though, that residential propinquity became less important in Paris than it had been in Bordeaux, because a closer examination of where family members lived reveals that all changes of address by the parents were limited to a mere five-hundred-metre radius. The sole exception to this was in the last months of their lives when they, along with Olinde and his family, moved to the adjoining eighth *arrondissement.* Just as revealing is that their other children nearly always set up residence within half a kilometer of the parents and of each other.

Residential propinquity suggests mutual aid. Though, of course, surviving sources do not allow us to gauge its nature and extent, we do know from the Pereire family papers just how much Émile Pereire gave to friends and family from Bordeaux (Ratcliffe 1972, pp. 200–201). We also know that he offered Isaac Rodrigues lucrative shares in the highly successful Paris-Saint-Germain railway company, while the inventory of the latter's estate that was made after his death in September 1846 shows that he kept them and that they made up no less than a third of his estimated wealth. We also know that three of Olinde Rodrigues's brothers-in-law (Andrade, Baud, and Sarchi) were, at one time or another, employed by the Pereires' railway enterprises. Most importantly, at least as far as Olinde Rodrigues is concerned, the Pereire brothers gave him financial help in the 1830s and 1840s, and this aid may well have enabled him to devote himself again to mathematics (he published a cluster of papers in the 1838–1840 period, which was precisely the moment when Émile Pereire first became wealthy) and to take up a series of other less successful projects in the 1840s. In his will Olinde Rodrigues asked his family never to forget that the Pereires' 'pocket-book has been mine, that they never understood that I could be in need when they themselves were not.'[9]

The combination of manuscript documents we have used, then, tells us at least something, though inevitably not nearly as much as we would like, about change and stability in Rodrigues's family life and that of others in their orbit. It helps us understand the values, support, and affection that sustained Olinde. A sign of the role these played is in the circumstances of his final, anguished, and bitter schism from the Saint-Simonian sect in February 1832. One reason for his leaving was that, unlike most of those still in the group at this time, he was married and had three young children. He also left because of pressure by his parents and from Émile Pereire, who had already broken with the sect the previous November, pressure that intensified after the police raided the sect at the end of January and there was the menace of legal proceedings against them. His departure was further precipitated by Enfantin's deliberate public revelation of (unnamed) indiscretions that Rodrigues's wife had apparently been cajoled into confidentially confiding to him (in the next chapter, Paola Ferruta attempts the difficult task of decoding the ambiguous written comments that his wife made about this). At its root, though, it was about family values. While Enfantin was groping for a new sexual morality and gender relations in and outside marriage, Rodrigues, in the speech he gave to explain why he was breaking with the sect, announced that, while he believed in divorce (not legal in France at the time, though it had briefly been permitted during the Revolution) and second marriages, he also believed in the sanctity of marriage

---

[9] *Archives Nationales. Minutier central. Étude* VIII, 1648. Dépôt judiciaire du testament de M. Olinde Rodrigues. Testament holographique, le 18 juillet 1847.

and the family unit (Rodrigues 1832*a*). In practice and in principle, family life remained a significant element throughout his life.

Faith

Not only did Olinde Rodrigues never practice Judaism, but there is nothing in surviving witness to suggest that he ever expressed concern about the issue of religion or was directly involved in the life of his community of origin. Examining his Jewishness nevertheless holds out a double promise. First, it was important in his life. If we are to understand him, we need to explain how it was that he was born into Judaism and was circumcised and yet not only did his family cease to be orthoprax and fail to bring him up in the faith, but he even married a Catholic and raised his children as Christians, all exceptional developments in the French Jewish community at the time. If we are to understand his career, we also need to explain how he was a victim of anti-Semitism in the early years of the Restoration for this closed off access to his chosen profession. Finally, if we are to fully comprehend his role in the Saint-Simonian movement, we need, if only briefly, to address the question of the place of Jews in Saint-Simonism because, *pace* some observers at the time and some scholars later, Jewishness was a significant element in the sect and the young Jews who were involved in it were a group that was marginal in relation to the centre that was the Jewish consistory, the governing body in community life, and yet they played a disproportionate role—as a catalyst—for intellectual and political change in French Jewry in the nineteenth century. This latter argument is part of the thesis presented by the Israeli scholar Michael Graetz (Graetz 1989), a translation of whose study was published in France in 1989 as *Les Juifs en France au XIXe siècle*. The title of the Hebrew original better indicates his thesis: *When the periphery invests the centre.*

Second, an examination of Rodrigues and his Jewishness promises a clearer understanding of the lived experience of acculturation in the Jewish élite in Paris at a moment of accelerating but never constant change and intellectual ferment. In view of the number of studies now available, it might well be asked, though, whether a better understanding of acculturation is still needed. We believe that it is. There are three principal reasons we might invoke to explain why. One is that the process is not only complex but emotionally charged, and the terms that are generally used for it are as much normative as descriptive. As a consequence, the study of Jewish acculturation is not so much a field as a minefield across which scholars must tiptoe at their peril. Acculturation, indeed, is always a fig leaf term that covers different degrees of adjustment that range from indifference to religious belief and practice (which rather than just loss of faith could at this time be a deliberate refusal to attend synagogue services whose forms some had come to regard as alien) to inobservance of the Sabbath, dietary laws, and holidays, to marriages with non-Jews, and to outright conversions to Christianity. Moreover, even the ultimate step that was apostasy could be one that was progressively reached or, in rarer cases, could be the result of genuine spiritual conversion. What complicates issues even more is that then, as now, these different degrees of acculturation were subject to varying degrees of reprobation within the Jewish community: ceasing to be orthoprax was one thing, while marrying a non-Jew and baptising children was quite another. Acculturation, then, needs to be further studied precisely because it is an emotion-laden and complex process. We believe, secondly, that much of the historiography of contemporary Jews had not always treated acculturation at this time with the empathy that the actors themselves deserve. While no scholar has followed the claim of Heinrich Graetz, one of the founders of Jewish historical studies, who claimed that assimilation and conversion were the result of an infection with the 'moral depravity' of gentile society (Graetz 1892, Vol. V, p. 448), there has been a long tradition that ascribes conversion to motives of self-interest and treats it only as an 'entry ticket' to European culture, as Henrich Heine, himself an apostate (and an unhappy one), ironically put it, or call it *Gefallsucht*, the need to be accepted in gentile society, as the large-scale conversions among wealthy Berlin Jews in the early nineteenth century were termed. A third argument might be invoked: we still know too little about the atmosphere and the debates in élite circles in the Paris of the last decade of the eighteenth century and the first half of the following century, as about the extent of the decline in religious observance, mixed marriages, and conversions. The author of a recent article (Landau 2002) has argued that during the entire nineteenth century only one percent of the Jewish population in Paris converted. Whatever the accuracy of this calculation and however comforting it might seem to defenders of orthopraxis, it drowns what was mainly, but not exclusively, an élite phenomenon in the mass of less prosperous native-born and even more numerous immigrant Jews. It therefore downplays the extent and significance of the phenomenon among the better-off.

While this is clearly not the place for any reassessment of acculturation and conversion, we might enter here three correctives that will help us better situate Olinde Rodrigues's and his family's abandonment of Judaism and the Jewish community. One is that, instead of merely stressing indifference and the desire to be accepted in gentile society, we should not overlook the enthusiasm and idealism that some felt for acculturation. These resulted from justifications that Jewish intellectuals had proffered in France and in Germany before the Revolution, from emancipation and the discourses and debates on political and civil society that flourished in the 1790s, and from the loosening of community controls, even after Napoleon reestablished them when he set up the consistories in 1808. Frochot, prefect of the Seine Department, wrote of Parisian Jews in a letter to the Minister of the Interior in June 1806:

> They have tasted the joy of having a fatherland; and for some of them the attachment they have for it has loosened the ties that bound them to their society of origin. From this stems the neglect of the religious cult, the disregard and even the disdain for different practices that were generally respected before the Revolution.[10]

Emblematic of the continuing enthusiasm for acculturation amongst at least some in the Parisian Jewish élite is the first Jewish journal to be published in the capital, *L'Israélite français* (1817–1818), edited by Mathis Dalmbert and Élie Halévy. Its very title, adopting, as it did, the term 'Israelite' rather than 'Jew' in the belief that the latter term had been contaminated by its association among gentiles with usury, signals their desire that coreligionists adopt the culture of wider society as did the journal's motto taken from Psalms XXVII 'hold to the country

---

[10]Letter from Adolphe to Gustave d'Eichthal, 27 August 1827. *Bibliothèque de l'Arsenal. Fonds Enfantin* 13746.

and preserve your faith.' It was up to Jews, the prospectus for the journal announced, 'to make new efforts in order the better to justify the sage legislation that had been adopted concerning them' (*L'Israélite français*, Vol. I, 1817, prospectus, pp. 214–220 and 237–242), and these were the arguments put forward in different contributions, including, incidentally, one by Myrtil Maas, Olinde Rodrigues's friend and collaborator at the time (*L'Israélite français*, Vol. I, 1817, pp. 301–307). A second corrective may be added. Though acculturation was far from being a new phenomenon (and it had gone furthest in the Bordeaux community), mixed marriages and conversions were both novel and still rare. In Bordeaux between 1793 and 1820 there had been thirty-six mixed marriages (Cavignac 1991, pp. 88–96) and in the capital twenty mixed marriages had already taken place by 1809, while there had been perhaps twenty-five by 1820 ((Kahn 1885, pp. 65–78) and (Piette 1983, p. 154)). They were also frowned upon in the community. Members of the Assembly of Notables that Napoleon convened in 1806 when he was trying to decide on a policy towards French Jews in general and Jewish money lending in Alsace in particular, were asked a series of (never innocent) questions about their attitudes towards wider society. In their responses, they were usually anxious to demonstrate to the emperor that they wanted and deserved to be citizens like any others. But on one question rabbis, more conservative than lay members but in a minority in the Assembly, had their way over their more assimilationist colleagues: on the issue of mixed marriages the Assembly declared it did not oppose them, as it had to since ceremonies were civil and required no authorization from religious authorities, but refused them any religious sanction. Conversions, too, were anathema to those who remained in Judaism because ceasing to be orthoprax was one thing, but going over to what many regarded as the other side, to the long-time persecutor, was held to be perfidy. When in 1823 David Drach, a rabbi and son-in-law of Chief Rabbi Deutz, infamously converted—it was, at least, a genuine spiritual conversion—his wife and children were secretly taken away to London though they were later reunited with him ((Ignace-Xavier Morel—who was a convert whose original name had been Lévy-Gumpel—1826) and (David Drach 1825, p. 291)). Such extreme measures were rare, but shunning and ostracism were not. Conversions also split families: when the d'Eichthal family converted to Catholicism in the early years of the Restoration, the father's in-laws in Nancy broke off relations, and even though his sister, Sophie, wife of a leading Prague banker, who remained orthoprax, did not break with them, she and her husband would later refuse to give permission for their daughter, Augusta von Laemel, to marry his son, Gustave (Ratcliffe 1977, p. 119). There is a final corrective: a mixed marriage or a conversion did not necessarily mean that a Jew reneged on his—or more rarely her—ethnic group. Jews were constantly made aware of their identity by the existence of poverty in their community and a deep-seated sense of obligation among the better-off to offer succour, by reminders that anti-Semitism was still present and by the fact that fellow Jews outside France were less free than they were (Albert 1982, pp. 249–274). We can give instances of this. The polyhistor, Olry Terquem, born in 1782 and, as was Olinde Rodrigues, interested in mathematics, cohabited with a Catholic and had three children with her before he married her, but he did not abandon his concern for his community: under the telling pseudonym of Tserphati (which means Frenchman), he wrote a long series of articles advocating reform of synagogue rituals (Landau 2001, pp. 169–187). Similarly, Joseph Salvador, born two years after Rodrigues to an already acculturated father and a Catholic mother in Montpellier, who at his own request would actually be buried in a Protestant cemetery, still devoted his life's work in Paris to defending the Jewish contribution to civilization. In his first published work (Salvador 1822), then, he set out to show that Moses had created a republic based on a social contract, the rule of law and equality, that made it a forerunner of the first French Republic. In the same study, moreover, he further argued that when Jesus was condemned to death, Jews had not been guilty of the deicide for which Christians had long criticized them. This and his later books served to reinforce pride in Jewishness in intellectual circles. It was greeted with enthusiasm in the Rodrigues family (Ratcliffe 1972, p. 198) and would be reissued in a much expanded version in 1828 (Salvador 1828).

It appears paradoxical that Isaac Rodrigues and his family would cease to be practicing Jews because for two decades, from 1806 to 1825, the father played a significant role defending, representing, and serving the community. In many ways this paradox is more apparent than real for we know that already in eighteenth-century Bordeaux Sephardim had shown scant concern about religious matters and their rabbis had not played a leadership role in running community affairs. We also know that, given the oligarchic manner in which the new consistorial system was organized which accorded power to the twenty-five leading notables in the capital who alone elected lay members of the consistory, control was vested in the élite and, above all, in the business élite, many of whom were now less orthoprax than the mass of the population they represented and served (Albert 1977, pp. 100–101). Isaac Rodrigues's role, though, was real, and it was threefold. His first was a public defence of French Jews, which he published in the *Revue Philosophique, Littéraire et Politique* in March 1806. During an economic recession, smallholders in rural Alsace were loudly complaining about the interest they had to pay on loans from Jewish money lenders. Just at the moment when these protests were brought to Napoleon's attention, the Catholic publicist and philosopher Vicomte de Bonald published a violently anti-Jewish article in the *Mercure de France* (8 February 1806) in which he argued that Jews would always remain an alien minority incapable of assimilating to French society. Rodrigues's reply was immediate and reveals much about its author. Not only was it cogently argued, for de Bonald's text was refuted point by point, but it invoked landmark works by Jewish thinkers and, above all, those published in debates on the present and future state of the Jewish community in France that were written in the decade preceding the outbreak of the Revolution. Rodrigues also showed he knew the works of Moses Mendelssohn because his article included a refutation of de Bonald's claim that the Jewish German philosopher had only modest talent (I. Rodrigues 1806). As do his publications in the first decades of the nineteenth century in the more recondite area of accountancy (I. Rodrigues 1804, 1810, 1811), his article reveals a high level of literacy. More importantly, though, it shows that he was cognizant of, and in agreement with, philosophical justifications for acculturation. He also made a second contribution to his ethnic group. In the weeks following the publication of Rodrigues's response to de Bonald, Napoleon set up the Assembly of Notables made up of representatives chosen by prefects from a list of candidates nominated by the different Jewish communities. At first a substitute, Rodrigues was named a full member of the Assembly as one of the representatives of the Seine department when one of the original nominees dropped out. It would have been useful if the minutes of the meetings of the Assembly, which were published at the time and have been recently reissued (Ayoun 1997), were more than laconic. Unfortunately, real discussions between what government officials at the time termed the philosophical party and the religious party took place outside the Assembly itself—and beyond the hearing of Napoleon's officials who attended formal sessions—and the official record is made up only of the formal answers to the questions that were posed to it. This means, then, that we cannot establish what role, if any, Isaac Rodrigues played in the deliberations. We do know, however, that the Assembly elected him one of its two secretaries, and we do know that this meant that he had to work closely with Abraham Furtado from Bordeaux, who was elected president and who was a powerful assimilationist (Malino 1978, pp. 65–111). Napoleon also summoned a second body representing Jews, the *Grand Sanhedrin*, which met for six weeks in February–March 1807. Only fifteen of its seventy-eight members were laymen—the others were rabbis—and Rodrigues was named to be one of them. Its recorded proceedings are as formal as those of the Assembly of Notables, but as important as its resolutions were to be in the subsequent history of French Jewry, these had been largely prepared by the Assembly of Notables, and there is less reason for us to lament the absence of fuller minutes of debates. There is more reason to regret that we cannot discover more about Isaac Rodrigues's third contribution to his community: his role as secretary of the Paris Consistory, a post he occupied for sixteen years from the body's first meeting in 1809. Surviving registers and correspondence files in the sadly depleted Consistory archives offer only glimpses of the role he played. It is clear, though, that he was at the heart of community affairs for a longer period than most others because, unlike regular members who were elected to serve limited terms, his was a modestly paid position to which he was named on a long-term basis.

From 1825 onwards, however, Isaac Rodrigues's name disappeared from both Consistory records and from the Ministry of Religion's well-kept files on nominations to the Electoral College and results of elections to the consistory. He played no further role in community affairs. We have found no direct evidence that would explain this sudden break, but there are indirect clues. One is that correspondence between Émile Pereire in Paris and his mother and friends in Bordeaux clearly indicates that the Rodrigues family had long ceased to attend synagogue and to observe dietary laws, the Sabbath, and Passover. Worse, Olinde, the eldest son, had married outside the faith in 1817, while Sophronie, the eldest daughter, and her husband, David Andrade, were already contemplating converting and baptizing their children. Though Olinde Rodrigues's mixed marriage had not provoked his father's resignation, it is conceivable that the threat of a more spectacular conversion by the son of the chief rabbi of Bordeaux might have led him to do so. When there were conversions in the families of more prominent community leaders, they usually felt constrained to resign their positions, as did Olry Worms de Romilly, who was president of the consistory, when, in 1843, his granddaughter married outside the faith. It is also conceivable that Isaac Rodrigues's stock-jobbing business—he was now in partnership with Émile Pereire, his dynamic son-in-law—took up more of his time and brought in more money and made his low-paying post at the Consistory less necessary. Whatever the precise motives for the break, until he did so, and as did others in his circle and the Jewish élite in general, he had proved more faithful to his ethnic origins than to the faith into which he had been born.

In order to understand why Olinde Rodrigues was not raised in Judaism and would marry outside it, we need, as a first step, to reconstruct his father's circle in the decades following his arrival in the capital. Any attempt at doing so has to recognize the central role played by the Fould banking house on rue Bergère in what is now the ninth *arrondissement*. The business, the family, and the building they occupied were the locus not only for the professional training of the young Jewish bankers who came there from Paris and, above all, from other Jewish banking families outside the capital and outside France, but also for the exchange of ideas about, and enthusiasm for, acculturation. Among them was Abraham Mendelssohn, the future founder of the Berlin bank of the same name, who liked modestly to claim that he had spent part of his life as just the son of the reputed German Jewish philosopher and the other part as just the father of his famous composer son (though it is revealing of his patriarchalism that he failed to mention his pianist and composer daughter Fanny). He did part of his banking apprenticeship at the Fould bank in 1802–1803. His intellectual sister, Dorothea, and Friedrich Schlegel, her *littérateur* husband, held a celebrated salon during their sojourn in the capital from 1802 to 1808, while Henriette, his younger sister, also had a salon in the building owned by Berr Léon Fould in the first decade of the new century. Their salons not only served as centres for intellectual exchange between Jews and non-Jews alike but undoubtedly for the direct transmission of the ideas of the Haskalah, that influential German movement, which in the late eighteenth century advocated Jewish acculturation to gentile society and in which their father, Moses Mendelssohn, was the most influential figure. Five out of six of the latter's children, we might remember, converted to Christianity (and Dorothea converted twice: she first adopted Lutheranism and later changed to Catholicism). Others, too, would pass through the Fould bank. Myrtil Maas, who would collaborate with Olinde Rodrigues in the following decade, worked there in 1817–1818, while Théodore Ratisbonne was there in 1818–1820. The latter was to go through a heartfelt religious conversion in 1827 (Ratisbonne described his state of mind when he converted in (Ratisbonne 1966, pp. 17–18, 116–118)), and his brother would follow him fifteen years later. Ratisbonne was to establish the famous or infamous, depending on the point of view, Catholic order of *Notre Dame de Sion*, whose principal mission was to convert Jews.

The Rodrigueses' circle of Jewish friends also included the Halévy family. The father, Élie Halévy, was a leading assimilationist, cantor at a Paris synagogue, and editor of the short-lived *L'Israélite français*. One of his two sons, Léon, would act as secretary to Saint-Simon in the last months of his life and collaborate closely with Olinde Rodrigues before and immediately after Saint-Simon's death. Both Léon and his composer brother Fromental would live in an apartment immediately above the Rodrigueses in the early 1830s. Though Fromental was to remain a practicing Jew, Léon married a Catholic late in 1832, and the couple had their son baptized at the beginning of 1834 (see (Halévy 1863, pp. 9–10) and Beatrice Philippe in (Loyrette 1996, pp. 54–65)). Halévy was not alone among those in the Rodrigues family circle in going so far as to abandon Judaism. The three children of the banker Louis d'Eichthal were baptized in 1817, and the parents followed them in 1820 (Ratcliffe 1977, pp. 113–118). Mathis Dalmbert, son of Simon Mayer Dalmbert, a leading figure in the consistory, and a graduate of the *École Polytechnique*, married his Catholic partner in 1831 and later baptized their three children born out of wedlock. As we have seen, in 1823 David Drach, at that time principal of the school run by the consistory and son-in-law of Chief Rabbi Emmanuel Deutz, abjured. Olinde Rodrigues married outside the faith in 1817, and his eldest sister and her husband had their children baptized in 1826. We do not know what motives lay behind the Andrades' conversion—available sources are mute on their decision—but, by looking at anti-Jewish sentiments that resurfaced with the return of the Bourbons in 1815, we are able to learn more about what may have lain behind Olinde Rodrigues's mixed marriage.

However sanguine assimilationists, such as Isaac Rodrigues, might be about the vistas that emancipation had opened up for them, whatever regime was in power, Jews in France still encountered anti-Jewish feelings that would always linger on in many corners of gentile society. The immensely popular German Jewish composer Giacomo Meyerbeer, who had changed his original name of Yaakov Liebmann Beer in 1815, believed he was successful despite his origins and bitterly but evocatively described the prejudice he had met in a letter he wrote in 1839 to Heinrich Heine (Letellier 1999):

> No pomade or bear grease, not even baptism, can grow back the foreskin of which we were robbed on the eighth day of life; those who, on the ninth day, do not bleed to death from this operation shall continue to bleed an entire lifetime, even after death.

Prejudice, of course, manifested itself in many different forms that ranged from the name-calling that Gustave d'Eichthal suffered at school (Ratcliffe 1977, p. 118) to outright discrimination. It was the latter which irrevocably changed Olinde Rodrigues's career. His first choice was to be a mathematician occupying a post in higher education. Unfortunately, the Bourbon Restoration gave more power and influence to the clerical party, and this was especially manifest in education. We might add that schools in nineteenth-century France were destined to remain a sensitive and contested issue on which church and secular views would clash, and Catholic clerics would sometimes resist the nomination of Jewish teachers, as they did most famously when the brilliant Isidore Cahen was appointed to teach the particularly sensitive subject of philosophy in a provincial *lycée* in 1849. The outcry was such that he resigned his position. Three decades earlier similar anti-Jewish prejudice and fears ended the hopes of academic careers of a cohort of outstanding young Jewish mathematicians.

We know, too, but we do not properly understand why, that, in contrast to other outstanding Jewish students in his age cohort in the last years of the Empire, Rodrigues did not secure admission to either the *École Polytechnique* or the *École Normale* (later renamed the *École Normale Supérieure*). We also know that Rodrigues failed to secure a regular academic post, despite the path-breaking research that earned him a doctorate in 1815. We shall better understand what happened to him if we know that similar fates befell others in the group of young Jewish mathematicians who had been trained at the *École Polytechnique* or the *École Normale*, where research and training in the discipline were especially remarkable at this moment. One of these was the polymath Olry Terquem, who was the originator of the term pedal curves, author in 1842 of the second analytical proof of the Feuerbach's theorem that the nine-point circle touches the incircle and the excircles of a triangle, and of two mathematics textbooks and editor of a periodical aimed at university-level mathematics students. In 1804 Terquem had been only the second Jew to have been admitted to the *École Polytechnique* and had begun his teaching career as a professor at the *Lycée Impérial* in Mainz, then under French control. He would lose this post, and the sole paid post he subsequently held was that of librarian at the Artillery School in Paris. This, though, did not prevent him from pursuing, and may have actually allowed him to pursue, his diverse intellectual activities. A second was Myrtil Maas (born 1796) and a third Serve-Dieu Abélard Lévy (born in November 1795), whose father was a Jewish immigrant from the Duchy of Modena who had begun as a *colporteur* (pedlar) and had married a Catholic in 1793. Both were admitted to the *École Normale* in 1813. After only two years there, the outstanding Lévy was made a tutor in mathematics at the school. Their careers, too, would be blocked by the return of the Bourbons. In a letter written in May 1816 to the Minister of the Interior the Paris Consistory complained that both men had been warned by officials at the Ministry of Education that they should seek careers outside academe because, they were told, the fact that they were Jewish would be of concern for some students' parents. As a result, Maas abandoned his academic plans and went, instead, to work as bookkeeper for a porcelain manufacturer (most likely for the firm run by the Jewish notable Baruch Weil, who had a factory at Fontainebleau, just outside Paris). He was lost to mathematics but was to have a highly successful career in the growing insurance industry in the capital. Lévy, on the other hand, persisted with his teaching ambitions but was only offered a post at the *lycée* at the tiny tropical colony of the *Île Bourbon* (now *Réunion*) in the Indian Ocean. Happily for mathematics (and mineralogy, in which he also had an interest), a storm forced the ship that was to take him there to seek shelter in an English port. He went ashore, decided to stay, and commenced his research career with British collaborators, only returning to Paris after the more liberal July Monarchy regime had been installed in France in 1830.

It is not impossible that this unofficial closing of the doors to academic careers had a role in Rodrigues's decision in 1817 to marry outside the faith and thereby be the first in his family to break the centuries-long tradition of endogamous unions, although the depth of his feelings towards his attractive wife must not be discounted. To seize the import of this, we need to know that Victorine Denise Martin, his bride, not only came from a family of very modest means but was also a devout Catholic and that their children would be baptized, raised as Christians, and that each would have a church wedding, one at the Madeleine and one at St. Louis d'Antin, both fashionable venues.[11] His decision to marry outside the faith may have been easier to make because others in his circle were already cohabiting with Christians and because not only was he not a practicing Jew but at no time did he even give voice to any need he might have felt for religious commitment. In this, his silence contrasts sharply with the mystical turn of mind of his brother Eugène, eleven years his junior, a spirituality which led him to stress the need for religious belief in letters he wrote (though probably in collaboration with others) as a Saint-Simonian convert in 1829. The sect published these as one of its key

---

[11]A number of clues suggest this. Rather than use her given forenames, she always preferred to be called Euphrasie, after Euphrasia, the Catholic patron saint of the day of her birth. The will that Olinde Rodrigues's widow drew up in November 1880 a few months before her death, which is to be found in the *Minutier Central* notarial records at the *Archives Nationales*, reveals that she remained pious: amongst other things, she made a bequest to the vicar of the Saint-François-de-Sales in the XVIIe *arrondissement*. We were allowed to consult her *Inventaire après décès* and attached documents, drawn up in July 1881 at the office of notary Patrice Philippot on the rue de la Chaussée d'Antin.

texts under the title *Lettres sur la religion et la politique*. In one passage there he would exclaim, 'You have to believe to be believed, you yourself have to burn to inflame others' (E. Rodrigues 1832, p. 13). Rather, Olinde shared this lack of religious commitment with others in his immediate circle, including, for instance, Benoît Fould, born in 1792 and son of Berr Léon, who declared at this very moment that he was absolutely indifferent to religion (Barbier 1991, pp. 77–78) and Émile Pereire (born 1800) who declared in a letter written in 1826: 'I am not religious... I was born into the Jewish faith. I remain in it because... I am firmly convinced that to change it would be necessary to be a sincere believer or a hypocrite' (Ratcliffe 1971, p. 1236). Whether or not Rodrigues's decision was a hypocritical one, we should also note that registration records at the Paris archives indicate that he was not alone in converting and that other branches of the family in Paris also converted in these same years: we know, then, that Edouard Rodrigues's new-born daughter was baptized in 1823 and that Hippolyte Rodrigues, from yet another branch, was already a Catholic when he married in a church ceremony in 1836. Even within Olinde's immediate family and apart from the conversion of his eldest sister and brother-in-law, two of his sisters would also later marry Christians (in 1827 and 1832, respectively).

Though he may not have married outside the faith for religious reasons and though his action was part of a pattern in the family, this does not mean the decision would have been taken without raising difficulties, particularly since in 1817 mixed marriages and conversions were more unusual than they would become. While available sources do not make us privy to any anguish and conflicts that Olinde Rodrigues's choice of partner likely generated, we do know that when the Andrades were talking of converting only a few years later, pressure was put on them not to abjure and that there were bitter discussions within the family. The Pereire correspondence also reveals that in the months immediately following Esther Anaïs Rodrigues's wedding with the Catholic Romain Regnault early in 1827 both Émile Pereire and his mother opposed the younger brother Isaac's proposed marriage to another sister, Rebecca Mélanie, insisting that, before there were any further alliances between the two families, the Rodrigueses undertake to permit no more marriages with Christians (Ratcliffe 1972, p. 199). No such promise proved forthcoming, and Isaac Pereire had to give up his love. He would later marry Laurence Lopes Fonseca, a cousin born and raised in the faith in Bordeaux. In the event, this would not change matters because neither their two children nor those from Pereire's second marriage to his niece Fanny Pereire were brought up in Judaism.

If Olinde Rodrigues's mixed marriage was not any result of a pressing desire on his part to adopt Catholicism for spiritual reasons and if, as we must suspect, his choice of bride still led to recriminations in family and entourage, why, then, did he make the choice he did? Allusions in the Saint-Simonian papers over a decade later tell us that his wife was a physically and intellectually attractive woman, and this suggests that her qualities were a factor. Choice could also have been prompted by Rodrigues's relations in gentile society: his bride was the daughter of a teacher at a Paris *lycée*, and these were the circles in which Rodrigues moved at this juncture. Other clues, though, suggest that his choice of bride may also have been a strategic one. One clue is that when, later in life, his other interests and the state of his finances allowed him to do so, he returned to his work in mathematics, which indicates how long lasting was his passion for the discipline. Another clue is that on his 1817 wedding certificate he described his occupation as 'doctor of mathematics.' Since the occupations of participants in civil ceremonies have always been self-attributed, the title he used may well be indicative of his continuing academic aspirations. If it is, we may further note that when in April 1821 his daughter's birth was registered, he still described himself as a 'doctor of mathematics,' suggesting he still hoped to be able to follow an academic career. In subsequent civil ceremonies he would describe himself as having a business occupation. There is a further clue: in the will he drew up towards the end of his life, he left what was, for the value of his estate, a major legacy of 10,000 francs to his widowed mother-in-law, saying that the money would ensure she would not lose her independence and that it was a solemn token of his gratitude to her. It is plausible that he was grateful to her for agreeing to allow her nineteen-year-old daughter to marry a Jew. It is at least possible, then, that he believed that marriage to a Christian would make him more acceptable to functionaries at the Ministry of Education for appointment to a teaching post. If this was indeed the case, he was wrong.

If his conversion to Catholicism was not a matter of faith, his conversion to the ideas of Saint-Simon was, and it was a faith he would keep throughout his life. His role in the Saint-Simonian movement and the fact that other young Parisian Jews also joined has long led some scholars to claim that their presence was important both in the elaboration of Saint-Simonian ideas and afterwards in applying so-called 'practical' Saint-Simonian theories in the business world and even in helping generate changed outlooks in the Jewish community itself. Since Olinde Rodrigues was the Jewish intellectual most involved for the longest time, we need to address this issue.

The myth that Jewish financiers and entrepreneurs would profit from their participation in the sect and from the ideas that it put forward to amass wealth and reinforce their power in key sectors of the French economy is an old and an anti-Semitic one. Already in the 1840s criticisms of the supposedly excessive power enjoyed by Jewish finance, and particularly by James de Rothschild, began to include criticisms of the disproportionate role that Jewish Saint-Simonians were purportedly playing in the economy. The most important of these attacks was published by the Fourierist Alphonse de Toussenel in 1845 (Toussenel 1845). Oft-repeated anti-Semitic myths of the power enjoyed by Jewish business in France, as elsewhere, have too often been shown to be erroneous to need discussion here (see (Landes 1998) for a recent restatement). The notion that Saint-Simonians, Jewish or gentile, found ideas, ideals, and solidarities that helped them make a major contribution to the French economy has also been shown to be mistaken ((Ratcliffe 1971) and especially (Ratcliffe 1995)). Of the Jewish Saint-Simonians, for example, only the Pereire brothers went on to remarkable careers. Gustave d'Eichthal, who came from a banking family, was prominent in the sect, but he never had a business career, and it was his brother Adolphe, who was never involved in it, who did. Charles Sarchi, another Jewish Saint-Simonian, who worked closely with Eugène Rodrigues, only had a modestly successful business career in subordinate positions, and it was his stockbroker brother Philippe, who had no connection with the group, who was the more successful. Olinde Rodrigues had already played a significant role in innovations in banking and insurance in the 1820s, but after his Saint-Simonian experience, he remained a marginal figure in Parisian finance, devoting as much of his energies as his finances and help from the Pereires permitted to promoting other causes closer to his heart.

A more recent thesis has also stressed the importance of Jewish involvement in Saint-Simonism: Michael Graetz's study of the evolution of the French Jewish community in the first two-thirds of the nineteenth century (Graetz 1989). The originality of his argument resides in his use of the concepts of centre and periphery as analytical tools to explain how French Jewry moved from introversion to assimilationism: small groups outside the conservative centre under the control of the rabbinate and the consistories, he argues, were responsible for bringing in new ideas and changing institutions. His argument has generally been greeted as innovative and suggestive (see, for instance, the endorsement—and application of the approach to the twentieth century—by Jacques Eladan (Eladan 1999–2000)). One of the suggestions Graetz makes is that the peripheral group that was the young Jewish Saint-Simonians was the first of these eccentric circles which had a disproportionate impact on the evolution of their wider community. If this is true, it would be an ironic twist in Olinde Rodrigues's life: he left Judaism only to impact it through the part he and others in his family and circle played in Saint-Simonism. We do not believe, however, that the claim is a valid one.

We do not believe this in part because the trouble with bipolar models is always that messy reality stubbornly refuses to conform to neat analytical categories, as it always has, for instance, for the developed/underdeveloped world dichotomy or for neo-Marxist theories about centre and periphery in the capitalist world system where some observers have found it necessary to advance the idea that there has always existed an intermediate category, the semiperiphery. We do not believe this, though, mainly because Graetz's analogy and argument do not work for Jewish Saint-Simonians. While it may be true that the official centre of the Jewish community in the capital was not an instrument of change, this stemmed from its structural and financial weakness, rather than from the conservatism of lay members of the consistory because, from the very beginning of its existence, many in the élite who dominated it were assimilationist (instead, conservatism was strongest in the mass of the population). More importantly, using texts written by Jewish Saint-Simonians between 1829 and 1831, Graetz argues that their authors were influenced by their contacts there to adopt a more favourable attitude to Jesus, the New Testament, and Christianity and that they found in the *Nouveau christianisme* (Saint-Simon 1825), as in Saint-Simonian historicism, a new pride in the contributions that ancient Jewish civilization had made to the evolution of humanity and stressed this in their own contributions to the sect's thinking, lectures, and publications.

The argument is open to criticism on a number of points. First of all, Jews who played a role in the sect were a tiny minority there. Even if we generously estimate the number of all those who took an active part in discussions in Paris at a hundred, Jews made up only one in twenty in the group. There were seven of them at the most: Olinde and Eugène Rodrigues, Émile and Isaac Pereire, Gustave d'Eichthal, Charles Sarchi, and Léon Halévy. All of them were from either Olinde Rodrigues's family or close circle. Though we cannot establish with certainty the contribution each made, we are able to say that Halévy left the group as early as 1826 and that Émile Pereire never subscribed to Saint-Simonian ideas on the need for a synthetic religion or to Enfantin's theories and was always on the fringes of the movement. Second, the implication of Graetz's text is that the Jews who joined were less assimilationist than they would become. In fact, by the time they became involved in Saint-Simonism, not one of the seven who became associated with the sect was orthoprax and all of them had long been assimilationist. Olinde Rodrigues, his brother, and Gustave d'Eichthal, indeed, were already outside the community. As we have pointed out, their pride in their ethnic groups had earlier been strengthened by Salvador's first publications, and in 1825 Léon Halévy had already written a history of ancient Jews that was strongly influenced by them (Halévy 1825). It may well be, indeed, that, rather than being the trigger of ethnic pride among Jewish Saint-Simonians, as Graetz has suggested, the favourable appreciation of the early Jews that is found in the *Nouveau christianisme* was mainly the result of the influence that Olinde Rodrigues and Halévy had on Saint-Simon. Third, finally, Graetz's literal reading of the texts written by Jewish Saint-Simonians is questionable. He attributes to them sole authorship when the texts were published under the aegis of the movement, and their purpose was to propagate its message and convert. It is more than likely that they were written after discussions in the group and issued after corrections by other members. A case in point is Eugène Rodrigues's *Lettres sur la religion et la politique, 1829* (E. Rodrigues 1831), a collection of letters he wrote in 1829 but was published after his death, which is one of the documents Graetz treats as emblematic. We have to admit that we have even fewer personal details about his brother than we do about Olinde himself. We do know, however, that, in the last months of its existence, the sect attributed influence and originality to his role in its discussions and to his published letters to two correspondents in the provinces that the Saint-Simonians had charged him with converting to the cause. We should bear in mind, though, that his premature death in January 1830, when he was not quite twenty-three, offered them a saintly figure—they buried him next to Saint-Simon's tomb in the Père-Lachaise cemetery—and the temptation to exaggerate his contribution, when, at least in the case of the content of his letters, it is likely that this was vetted in the group before the missives were sent. Whether or not this was the case, there is a second problem with Graetz's analysis of the text. He concludes that Eugène Rodrigues gave more space to the historical importance of Jews, Moses, and the ancient Israelites than did other Saint-Simonian texts, when we know that the *Nouveau christianisme*, one of their foundational texts, already did so. He also argues that the publication in the same volume of a French translation of Lessings' *Erziehung des Menschengeslechts* (*The education of the human race*), originally issued in 1780, shows that Rodrigues's purpose was both to emphasize the historical importance of Judaism and to stress toleration. In truth, Lessings' importance for the sect was that, as they did, he had stressed the need for a new synthetic religion that would supersede both Judaism and Christianity. Thus, while we should not gloss over the pride that Jewish Saint-Simonians showed for the achievements of their ancestors—Olinde Rodrigues himself liked to name Moses among history's great social reformers, as Graetz rightly reminds us—it does not seem tenable to maintain they came to Saint-Simonism because it accorded a greater place in history to Jews, that they discovered there a pride in their past, or that their role in a radical gentile movement was a significant catalyst for change in the Jewish community.

The attitudes towards their Jewish present and past that Olinde Rodrigues, his family, and his circle adopted tell us something about the mentalities and experience of the emancipation generation, about the hopes, fears, and contradictions of acculturation. That Rodrigues and a handful of other intimates joined the Saint-Simonian sect tells us something of the influence that the young Olinde Rodrigues had on his immediate circle, as about their aspirations for a new society, but less about their Jewishness and Judaism.

Conclusions

In this chapter we have argued that one way to secure a better understanding of Olinde Rodrigues is to insert his experience into the context of a family and immediate circle living under conditions of accelerating but always inconstant change. The approach we have adopted and the patchwork of manuscript and published sources we have interrogated in our effort to make this insertion are a pis aller. As such, they cannot compensate for the feebleness of his first-person voice and the paucity of revealing testimony on him from intimates. They possess, then, no 'open sesame' quality that affords us entry to the inaccessible: they permit us to seize the outer world of acts and gestures more than they do the inner world of attitudes and motives. However, combining information from serial sources, data from civil and religious registration, and more intimate details gleaned from notarial records, with a critical reading of sparse published texts by family and friends, and a reassessment of what is known about patterns of thinking and behaviour in the wider context of Jewish communities, gentile bourgeoisie, and the burgeoning city that was Paris, allows us to sneak up on our subjects especially at vital junctures when they unknowingly reveal themselves as individuals but also as products of their times. We therefore understand a little better how, in the space of just over a half century between Isaac Rodrigues's arrival in Paris during the First Republic and Olinde Rodrigues's death in the twilight of the Second, the family moved from the relatively liberal but still orthoprax (at least in outward forms) Sephardic community in Bordeaux, where life was still centred on family and ethnic group and occupations generally perpetuated across generations, to the more diverse and more dynamic environment of Paris. In the process, some family patterns changed but others continued, and as it did in society as a whole, family still offered crucial social capital for its members, including, as we have seen, Olinde Rodrigues. He and his family ceased to practice Judaism and even left the community. Their experience was not that of the vast majority of the rapidly increasing numbers of Jews in the capital, and it was not even that of all members of their circle—the Foulds and the Pereires did not abjure, for instance, and nor did Myrtil Maas, who would later serve on the Paris consistory—but the issues of faith and acculturation were not only earnestly discussed and defended in their circle, but a distinct if multifaceted pattern of acculturation developed. Olinde Rodrigues lost a sectarian faith and found new faith in a sect that promised a future earthly world without class divisions and prejudice and with a common religion. At the same time, he made significant contributions to wider society, to mathematics, to Saint-Simonism, and to banking and finance. His, his family's, and his circle's lived experience over five decades would have been inconceivable only a few short years before his birth.

Bibliography

Albert, P. C. (1977). *The modernization of French Jewry: Consistory and community in the nineteenth century.* Brandeis University Press.

Albert, P. C. (1982). Ethnicity and Jewish solidarity in nineteenth-century France. In *Mystics, philosophers and politicians. Essays in Jewish intellectual history in honor of Alexander Altmann* (Yehuda Reinharz and Daniel Swetchinski, eds.), pp. 249–274. Duke University Press, Durham, NC.

Ayoun, R. (1997). *Les juifs de France de l'émancipation à l'intégration 1787–1812.* L'Harmattan, Paris.

Barbier, F. (1991). *Finance et politique. La dynastie des Fould XVIIIe–XXe siècle.* Amand Colin, Paris.

Berkovitz, J. R. (1989). *The shaping of Jewish identity in the nineteenth-century France.* Wayne State University Press, Detroit.

Bigo, R. (1947). *Les banques françaises au cours du XIXe siècle. L'avènement du crédit, sa distribution et ses instruments.* Recueil Sirey, Paris.

Blamont, C. (2002). Généalogie d'un fils de colporteur [le savant Abeillard Lévy...]. *Revue du Cercle de Généalogie Juive*, **18**, 20–23.

Bottin. S. (1808–1850). *Almanach du commerce de Paris et des départements.* Bureau de l'Almanach du Commerce, Paris. [Titles vary; first published by J. de La Tynna.]

Cavignac, J. (1987). *Dictionnaire du judaïsme bordelais aux XVIIIe et XIXe siècles: biographies, généalogies, professions, institutions.* Archives départementales de la Gironde, Bordeaux.

Cavignac, J. (1991). *Les Israélites bordelais de 1780 à 1850: autour de l'émancipation.* Publisud, Paris.

Cohen, D. (1980). *La promotion des Juifs en France à l'époque du Second Empire (1852–1870)* (2 vols.). H. Champion, Paris.

Coleman, J. S. (1988). Social capital in the creation of human capital. *American Journal of Sociology*, **94**, supplement, 95–120.

Coleman, J. S. (1990). *Foundations of social theory.* Belknap Press, Cambridge, MA.

Courteault, P. (1925). Un bordelais saint-simonien. *Revue Philomatique de Bordeaux et du Sud-Ouest*, **28**, 152–166.

Daumard, A. (ed.) (1973). *Les fortunes françaises au XIXe siècle: enquête sur la répartition et la composition des capitaux privés á Paris, Lyon, Lille, Bordeaux et Toulouse.* Mouton, Paris.

Daumard, A. (1996). *La bourgeoisie parisienne de 1815 à 1848.* Albin Michel, Paris. [First published 1963.]

Drach, D.-P.-L.-B. (1825). *Lettre d'un rabbin converti aux Israélites ses frères sur les motifs de sa conversion.* Beaucé-Rusant, Paris.

Drach, D.-P.-L.-B. (1827). *Deuxième lettre d'un rabbin converti aux Israélites ses frères sur les motifs de sa conversion. Les prophéties expliquées par les traditions de la Synagogue.* Published by the author, Paris.

Drach, D.-P.-L.-B. (1844). *De l'harmonie entre l'église et la synagogue ou perpétuité et catholicité de la religion chrétienne* (2 vols.). P. Mellier, Paris.

Eladan, J. (1999–2000). Intellectuels périphériques et communauté structurée. *Cahiers du centre d'études et de recherche sur l'identité Juive.* 8. [Electronic version, no pagination.]

Gille, B. (1959). *La banque et le crédit en France de 1815 à 1848.* Presses Universitaires de France, Paris.

Girard, A. (1981). *Le choix du conjoint. Une enquête psycho-sociologique.* Presses Universitaires de France, Paris.

Graetz, H. (1891–1896). *History of the Jews from the earliest times to the present day* (6 vols.). Jewish Publication Society of America, Philadelphia.

Graetz, M. (1989). *Les Juifs en France au XIXe siècle. De la Révolution française à l'Alliance Israélite universelle.* Seuil, Paris.

Halévy, L. (1825). *Résumé de l'histoire des Juifs anciens.* Lecointe et Durey, Paris.

Halévy, L. (1863). *F. Halévy, sa vie et ses œuvres, récits et impressions personnelles, simples souvenirs.* Hengel, Paris.

Hubbard, G. (1857). *Saint-Simon, sa vie et ses travaux.* Guillemin, Paris.

Jordan, R. (1994). *Fromental Halévy. His life and music, 1799–1862.* Kahn and Averill, London.

Kahn, L. (1885). Les professions manuelles et les institutions de patronage. *Histoire de la communauté Israélite de Paris*, Vol. III. Librarie A. Durlacher, Paris.

Kohn, J.-L. (1994). *La bourgeoisie juive à Paris au Second Empire*. Unpublished Ph.D. thesis. Université de Paris.

Landes, David S. (1998). *The wealth and poverty of nations: why some are so rich and others so poor*. Norton, New York.

Landau, P.-E. (2001). Olry Terquem (1782–1862). Régénérer les juifs et transformer le judaïsme. *Revue des Études Juives*, **160**, 169–187.

Landau, P.-E. (2002). Se convertir à Paris au XIXe siècle. *Archives Juives*, **35** (1), 27–43.

Letellier, R. I. (1999). *The diaries of Giacomo Meyerbeer*. Vol. I. Fairleigh Dickinson University Press, Madison, NJ.

Lévy-Leboyer, M. (1964). *Les banques européennes et l'industrialisation internationale dans la première moitié du XIXe siècle*. Presses Universitaires de France, Paris.

Lifshitz-Krams, A. (1996). *Mariages religieux juifs à Paris 1848–1872*. Cercle de Généalogie Juive, Paris.

L'Israëlite français. Ouvrage moral et littéraire (2 vols.) (1817–1818).

Locci, J.-P. (1988). *Fonderies et fondeurs: histoire des établissements métallurgiques du Vaucluse aux XIXe et XXe siècles*. Association pour la sauvegarde et la promotion du patrimoine industrielle en Vaucluse, Avignon.

Locke, R.P. (1992). *Les saint-simoniens et la musique*. Mardaga, Paris.

Loyrette, H. (ed.) (1996). *Entre le théâtre et l'histoire. La famille Halévy (1760–1960)*. Librarie Arthème Fayard, Paris.

Malino, F. (1978). *The Sephardic Jews of Bordeaux: Assimilation and emancipation in Revolutionary and Napoleonic France*. University of Alabama Press.

Manuel, F. E. (1962). *The prophets of Paris*. Harvard University Press, Cambridge, MA.

Morel, I.-X. (1826). *Renseignements relatifs à la persécution dont M. Drach, rabbin converti, a été l'objet*. Imprimerie de Gueffier, Paris.

Nahon, G. (2003). *Juifs et judaïsme à Bordeaux*. Mollat, Paris.

Neher-Bernheim, R. (1983). Un savant juif engagé: Jacob Rodrigues Pereire. *Revue des Études Juives*, **152**, 373–451.

Œuvres de Saint-Simon et d'Enfantin... publiées par les membres du conseil institué par Enfantin pour l'exécution de ses dernières volontés. (1865–1878) (47 vols.). E. Dentu, Paris.

Oukhemanou, A. (1998). La communauté juive de Bayonne-Saint-Esprit d'après les registres de 1808 (aspects démographiques et socio-économiques). *Revue des Études Juives*, **157**, 151–167.

Pereire, E. (1860). *Tables de l'intérêt composé des annuités et des rentes viagères*. Imprimerie Dupont, Paris.

Piette, C. (1983). *Les Juifs de Paris (1808–1840). La marche vers l'assimilation*. Presses de l'Université Laval, Québec.

Piette, C. and B. M. Ratcliffe. (1993). Les migrants et la ville: un nouveau regard sur le Paris de la première moitié du XIXe siècle. *Annales de Démographie Historique*, pp. 263–302.

Ratcliffe, B. M. (1971). Les Pereire et le saint-simonisme. *Economies et Sociétés*, **5**, 1215–1255.

Ratcliffe, B. M. (1972). Some Jewish problems in the early careers of Émile and Isaac Pereire. *Jewish Social Studies*, **34**, 189–206.

Ratcliffe, B. M. (1975*a*). Some banking ideas in France in the 1830s: The writings of Émile and Isaac Pereire. *Revue Internationale d'Histoire de la Banque*, **10**, 23–47.

Ratcliffe, B. M. (1975*b*). Crisis and identity: Gustave d'Eichthal and Judaism in the emancipation period. *Jewish Social Studies*, **37**, 122–141.

Ratcliffe, B. M. (1976). Saint-Simonianism and messianism: the case of Gustave d'Eichthal. *French Historical Studies*, **10**, 484–502.

Ratcliffe, B. M. (1977). Gustave d'Eichthal: An intellectual portrait. In *A French sociologist looks at Britain. Gustave d'Eichthal and British society in 1828* (B. M. Ratcliffe and W. H. Chaloner, eds.), pp. 109–161. Manchester University Press.

Ratcliffe, B. M. (1991). Classes laborieuses et classes dangereuses à Paris pendant la première moitié du XIXe siècle?: The Chevalier thesis reexamined. *French Historical Studies*, **17**, 542–574.

Ratcliffe, B. M. (1995). The Saint-Simonians in the French economy: Toward an understanding of ideas, ideals and action. In *Fra spazio e tempo* (Ilaria Zilli, ed.) (3 vols.). Vol. 2, pp. 733–754. Edizioni Scientifiche Italiane, Naples.

Ratisbonne, T. (1966). Mes souvenirs. Les Presses monastiques, Saint-Léger-Vaubon.

Reinach, S. (1901). Note sur une famille juive de Novellara (Italie). *Revue des Études Juives*, **43**, 135.

Rodrigues, E. (1832). *Lettres sur la religion et la politique, 1829.* Au Bureau de l'Organisateur, Paris.

Rodrigues, H. (1893). *Papiers de famille, afin de préserver ma mémoire et celle de mon fils Edgar de toute compromission, de toute négligence, à propos des apostasies de le rue Spontini.* Imprimerie Vve P. Larousse, Paris.

Rodrigues, I. (1804). *Observations sommaires et critiques, sur la méthode simplifiée de la tenue des livres de E. T. Jones.* Imprimerie A. Bailleul, Paris.

Rodrigues, I. (1806). *Observations sur un article de M. de Bonald, sur les Jufs, inséré dans le 'Mercure de France' du 8 février 1806. Extrait de la 'Revue Philosophique, Littéraire et Politique' numéro 9, 21 mars 1806.* n.d.n.p.

Rodrigues, I. (1810). *La tenue des livres théorique et pratique ou nouvelle méthode d'enseignement.* André Racle, Bordeaux.

Rodrigues. I. (1811). *Observations adressées à M. Le Conseiller d'Etat directeur général de l'imprimerie et de le librairie.* Béaume, Paris.

Rodrigues, O. (1815). *De l'attraction des sphéroïdes. Thèse soutenue devant la Faculté des sciences de Paris, le 28 juin 1815.* n.d.n.p.

Rodrigues, O. and M. Maas. (1820). *Théorie de la Caisse hypothécaire ou examen du sort des emprunteurs, des porteurs d'obligations et des actionnaires de cet établissement.* Delaunay, Paris.

Rodrigues, O. (1825). De l'industrie. Considérations générales sur les banquiers. In H. de Saint-Simon, L. Halévy, and O. Rodrigues, *Opinions littéraires, philosophiques et industrielles.* Bossange Père, Paris. pp. 161–199.

Rodrigues, O. (ed). (1830). *Doctrine de Saint-Simon.* Au Bureau de l'Organisateur, Paris.

Rodrigues, O. (1832*a*). *Aux Saint-Simoniens [13 février 1832]. Bases de la loi morale proposée à l'acceptation des femmes.* Imprimerie d'Everat, Paris.

Rodrigues, O. (1832*b*). *Saint-Simon, son premier écrit: Lettre d'un habitant de Genève à ses contemporains, 1820; sa parabole politique, 1819; le Nouveau christianisme, 1825.* Librairie Saint-Simonienne, Paris.

Rodrigues, O. (1840*a*). *Les Peuples et les diplomates. La Paix ou la guerre. [Signé: Olinde Rodrigues, 18 septembre 1840].* Imprimerie de Schneider et Legrand, Paris.

Rodrigues, O. (1840*b*). *La Paix et le travail. Journal Politique et Litéraire.* [Pamphlet, n.p.n.d.]

Rodrigues, O. (1840*c*). *De l'organisation des banques à propos du projet de loi sur la Banque de France.* H. Fournier, Paris.

Rodrigues, O. (1841). *Poésies sociales des ouvriers, réunies et publiées par Olinde Rodrigues.* Paulin, Paris.

Rodrigues, O. (1848*a*). *De l'organisation du suffrage universel, proposition d'un nouveau mode électoral.* Saint-Jorre, Paris.

Rodrigues, O. (1848*b*). *Projet de constitution populaire pour la République française.* Chaix, Paris.

Rodrigues, O. (1848*c*). *Théorie des banques.* Chaix, Paris.

Saint-Simon, H. de (1825). *Nouveau christianisme.* Bossange Père, Paris.

Saint-Simon, H. de, L. Halévy, and O. Rodrigues. (1825). *Opinions littéraires, philosophiques et industrielles.* Bossange Père, Paris.

Salvador, J. (1822). *Loi de Moïse ou système religieux et politique des hébreux.* Ridan, Paris.

Salvador, J. (1828). *Histoire des institutions de Moïse et du peuple hébreu* (3 vols.). Ponthieu, Paris.

Schwartzfuchs, S. (1989). *Du Juif à l'Israélite. Histoire d'une mutation, 1770–1870.* Fayard, Paris.

Silvera, A. (1966). *Daniel Halévy and his times, a gentleman-commoner in the Third Republic.* Cornell University Press, Ithaca, NY.

Simon-Nahum, P. (1988–1989). *Contribution à l'étude de la bourgeoisie intellectuelle juive parisienne, 1830–1914.* Unpublished Ph.D. thesis, E. H. E. S. S., Paris.

Toussenel, A. de (1845). *Les Juifs, rois de l'époque, histoire de la féodalité financière.* Librairie de l'Ecole Sociétaire, Paris.

Vinçard, J. (1878). *Mémoires épisodiques d'un vieux chansonnier saint-simonien.* Dentu, Paris.

CHAPTER 4

Euphrasie and Olinde Rodrigues: The 'Woman Question' within Saint-Simonism

Paola Ferruta

There were two questions crucial to Saint-Simonian doctrine that had a profound effect on Olinde Rodrigues's position within the community, since his views were opposed to those of Enfantin. One was the status of women in the community and in society, and the other was the redefinition of sexual morality. These critical questions, especially the second one, created tensions in the movement that eventually resulted in its demise. The study of those documents that exist is thus important in order to understand the stresses that had such a profound effect on Rodrigues's life. Rodrigues was a staunch family man, and a study of the life of his wife Euphrasie will throw light not only on the personality of the woman who followed and supported Olinde in his political and religious undertakings but also on the conditions under which Olinde had to pursue his intense and varied interests. The account of Euphrasie's experiences from 1830 to 1832 will greatly help clarify the historical process which led the Saint-Simonian community to its end, one of the most critical episodes in Rodrigues's life. Euphrasie's reaction to the crisis represents a refutation of Enfantin's male autocracy, as well as a rejection of the mere political consciousness achieved by some other female members of the movement who were rendered incapable of opposing him or even altering his ways. Euphrasie's intuition was that a mere political emancipation of women would not be sufficient to lead them to a state of autonomous and independent self-confidence.

As already mentioned before in this book, the study of Olinde Rodrigues's life is hampered by the lack of documentary evidence directly provenanced from him, beyond, of course, his printed works. This chapter will be mainly based on three extant letters from Rodrigues's circle, one of them from his wife and the other two from fellow Saint-Simonians addressed to him, which will help to understand the utopian milieu that he partly created but with which he had to contend.[1]

[1] The collection of Saint-Simonian manuscripts and books at the *Bibliothèque de l'Arsenal* in Paris cannot be regarded as a 'Saint-Simonian Archive' in the true sense of the word, since there is no precise classification of all the manuscripts, which makes it difficult to find certain items that are not evidently part of the Saint-Simonian correspondence. The archive, in fact, consists only of a number of *Fonds*, each one named after single members of the movement. Not all of them are kept in the Arsenal, and it is striking that none of the *Fonds* is named after Rodrigues. An important part of the documents regarding him is found in the main *Fonds* under the name of *Enfantin*, which mainly contain letters or duplicates of letters selected or copied by Prosper Enfantin and his followers, and there is no doubt that some parts of this selection were done, mainly by Enfantin, with a view to present himself to posterity, as well as the history of the Saint-Simonian movement, in a way that suited him. As will be shown in this article, Rodrigues's opinions diverged profoundly from those of Enfantin, so that in the *Fonds Enfantin* the name of

Birth and marriage

Who was Euphrasie Rodrigues? A visit to the *Archives de Paris* in the outskirts of the capital gives surprising results. Although most of the civil registration documents were burned in 1871, it is possible to read the original certificate of her marriage, a document of several pages. Even her birth certificate was registered a second time after 1871, and there is a duplicate. For several other members of the Rodrigues family their dates of birth and death and the names of their wives or husbands are recorded with similar accuracy. This means that these records cover many of the couples or single individuals who participated in some way in the Saint-Simonian movement. The Saint-Simonian urge to leave tracks, to preserve the memory of events and of people, reflects the particular attention paid by the Rodrigues family to their past and traditions. This similarity should not be interpreted as accidental or as a simple repetition of a previous original pattern.

Thanks to these two certificates, Euphrasie's birthday and date of marriage can be exactly established. She was born on 13 March 1798. When she was in her late teens and already an accomplished pianist, Benjamin Olinde Rodrigues, himself passionate about music, fell deeply in love with the young musician and soon married her, thus abandoning his religion of origin.[2] They married in Paris when she was nineteen, the bride giving her name as Victorine Denise Martin. A brief speculation about the reasons for her adopted name of Euphrasie would still leave the question open. We surely cannot assume that it was an entirely voluntary, personal choice. The documents show that her second name, Denise, was also her mother's name.[3] A historical digression about the name Euphrasie may be helpful in illustrating the religious atmosphere prevailing at the beginning of the nineteenth century and in particular her family's personal taste. The name sounds exotic, with a subtle oriental aura, and a glance at the *Acta Sanctorum* and at the *Vie de Sainte Euphrasie* tells us the story of the lives of Euphrasie and her mother in the fourth century AD at the court of Emperor Theodosius the Young in Byzantium. These two women had the same name. They secretly left the town after the death of Euphrasie's father jointly to enter a monastery in Egypt, famous for its austerity and peculiar rules. Other sources report that the daughter had to disguise herself as a man in order to hide from the soldiers. The young Euphrasie died in Egypt near Alexandria on 13 March AD 410.

Several works on the life of Sainte Euphrasie describe the celebrations and processions that were organised in France during the nineteenth century on every

Rodrigues appears only when it is indispensable, and his personality is never brought to light in all its facets. There are written comments on the top and side of the manuscripts, as in a Holy Scripture. We can recognize Enfantin's personal notes in his own handwriting: he signed himself PE. A part of the *Archives* were copied in 1822–1833, during the months of his imprisonment in *Sainte-Pélagie*. The handwriting changes from time to time, and Enfantin was not the only one copying. In using the material in the Saint-Simonian *Archives*, it must be remembered that they are presented in a deliberately selective way. They are not just a casual jumble of papers.

[2]'Extremely passionate about music, he fell in love with a young woman, excellent musician, and despite the difference in religion (she was Catholic), he married her.' This is the testimony, dated 14 April 1883, of a close associate of Rodrigues, Gustave d'Eichthal. (Note sur les saint-simoniens, *Fonds d'Eichthal, Bibliothèque Dosne-Thiers*, Carton IV, cited by Hervé le Bret in *Les Frères d'Eichthal: Gustave, penseur et saint-simonien, et Adolphe, homme d'action DEA d'Histoire Moderne et Contemporaire*, Université de Paris IV Sorbonne [unpublished], p. 182.)

[3]The French are in the habit of giving daughters their mother's name as second name; a son usually has his father's name as second name, but for Jews this is forbidden.

13, 14, and 15 of March publicly to honour the memory of this saint. We may note the connection between the date of birth of Victorine Denise and Sainte Euphrasie's feast day: the reason for her having adopted this name might have been her mother's inclination or Euphrasie's own personal choice, or it could have been a sort of battle-name that suited her militancy within the Saint-Simonian movement. It was perhaps invented by her husband, as a symbol of their intimacy in a 'new life' together.[4] Or it could be an allusion to the Egyptian or oriental fashion of that time, particularly in vogue amongst Saint-Simonians.

The beginnings of Saint-Simonism and the 'woman question'

The early years of the July Monarchy established in 1830 were marked by a social, religious, and political awakening. There were signs of a new consciousness, exacerbated among French industrial workers. The urban bourgeoisie faced the emergence of social problems caused by the spread of industrialization and the rapid growth of the cities in the preceding decade. Henri de Saint-Simon had indeed been a 'crisis' philosopher, drawing plans for the reconstruction of the world after the French Revolution and the Napoleonic wars. Frank Manuel (Manuel 1956) points out that he was a 'man of intuitions,' requiring his disciples not only to develop an intellectually complex body of doctrine but to prepare for action, although his views were never completely formulated. Saint-Simon may be considered as the 'aristocratic messiah' of the restructuring of society: in his schemes of social reorganization he conceived the governance of France by an élite of scientists and industrialists, denouncing the parasitism of the aristocracy. As shown in Chapter 2, Saint-Simon's followers constituted themselves as a church in 1829, four years after his death, eventually under the charismatic leadership of Prosper Enfantin, who displaced Rodrigues.

Charles Bouglé (Bouglé 1918, p. 56) emphasizes that the 'woman question' is the central problem of Saint-Simonism, and he observes that it became the 'collective obsession' of its adherents, adding that the Saint-Simonian 'rehabilitation of the flesh' discredited socialism for a long time. Saint-Simonian theories hinge on these two problems, emancipation of women and sexual freedom, which they take to be interrelated, and they declare the proletariat and women to be the two oppressed classes.

Olinde Rodrigues recorded a sentence that Saint-Simon spoke during one of their last conversations: 'man and woman: that is the social individual.' These words were destined to be significant in the debates of the Saint-Simonian school. The antithesis between the then-current military order and the pacific order of the future world was founded on the claim for women's political rights. Feminism and pacifism have the same meaning in Saint-Simonian terminology. The concept of the social individual as an androgynous being exists also in Charles Fourier's work, and this formula opened several new perspectives: Auguste Comte accepted it, as did Proudhon, especially when the 'hermaphrodite' becomes the administrator of justice (Bouglé 1918, p. 66). Neither the first nor the second, however, was interested in women's emancipation.

[4]His death certificate shows she was still beside him in their Parisian flat (rue d'Amsterdam is in the present 8*ème arrondissement*) when he died in 1851.

In later Saint-Simonian doctrine a bipolar entity, the sacerdotal couple, formed the complete social individual. The complementary union between the female 'passive, realizing' half and the male 'active, conceiving' half was first understood on a divine and then on a human level. The concept of the 'holy couple' is central to Saint-Simonian political philosophy. There still is no study of this aspect of the doctrine, a reelaboration of mystical ideas circulating in Europe at that time and linked to the great romantic myth of the female messiah, mediator between God and the new Adam, whose mission was to civilize the world.

It was under the influence of Enfantin's powerful charisma that the majority of the Saint-Simonians began to espouse ideas concerning the position of women and relations between the sexes, as well as the 'rehabilitation of the flesh.' Enfantin grounded his proposals for the destruction of the 'law of faithfulness' on his distinction between 'constant' and 'mobile' natures, the latter having to follow their instincts to satisfy their desires, if need be by adultery or prostitution; the struggle for the liberation of women was maintained to be proportional to their liberation of the 'law of faithfulness.' These proposals were offensive to Saint-Amand Bazard[5] and a number of others, but Enfantin vehemently denounced Bazard's authoritarian attitude as well as his incapacity to resolve the 'woman question'; as a result, on 11 November 1831, Bazard announced his decision to leave the church. This was fatal to the Saint-Simonian religion.[6]

Bazard's departure was followed by Rodrigues's alleged 'apostasy,'[7] when he declared himself the true repository of the doctrine and the only true heir of Saint-Simon, in order to protect the Saint-Simonians from the derision that fell upon them.[8] Rodrigues left the group together with other disciples, so that the crisis within the Saint-Simonian church came to a head. Bazard and Rodrigues expressed

---

[5]Bazard was employed at the customs office, and he was a hero of the defence of Paris in 1814; he was a free-mason, cofounder of the *Charbonnerie* in France in which he held a high rank. He was a convinced republican rationalist; after the failure of the Belfort conspiracy, he abandoned subversion and in 1825 joined the Saint-Simonian team of the *Producteur*. Responsible for the *Exposition de la doctrine saint-simonienne*, he was elected at the end of 1829, together with Enfantin, *Chef suprême de la religion saint-simonienne.*

[6]Bazard was born out of an adulterous affair by his mother, which may have made him more sensitive to the implications of promiscuity. His father had refused to see him until he met his wife, the vivacious and aimable Claire, but after that he accepted the family and his son. Claire Bazard was the mediator between Saint-Amand and the world until his death, being very successful on a social level. She was the most influential Saint-Simonian woman and the only one who had access to the inner circle around Enfantin, where the *doctrine saint-simonienne* was conceived by the few.

[7]He claimed for himself his master's heritage, excluding the others, and maintained that he was the only true repository of Saint-Simonian doctrine. He published alone the first volume of Saint-Simon's work, with his own preface. He wanted to issue another ten volumes, but he never succeeded.

[8]There is no doubt that Rodrigues was always interested in practical, social, and economic problems and did not go in for grand prospects à la Enfantin. A merely abstract intellectual activity was never his aim; even his mathematical works and discoveries are mostly linked to practical applications. In putting intellectual activity into contact with material realizations, he was a more faithful follower of Saint-Simon than Enfantin. We must not forget either that the clue for understanding the original contribution of all Saint-Simonians is to interpret religion as 'society.' *Religio* is what binds humanity together as one sole body. In this respect we can understand Rodrigues's engagement as 'alter ego' of the New Messiah Saint-Simon.

very different theoretical reservations,[9] while Enfantin and his followers put their antagonists' private lives under scrutiny.

Olinde Rodrigues's place within the Saint-Simonian community will be illuminated through that of his wife, which we shall do by discussing three extant letters that illustrate the position they were expected to have as a Saint-Simonian sacerdotal couple, as well as the assumed role of the female messiah.[10] The first is by Michel Chevalier, the second by Edmond Talabot, both Saint-Simonians, and the third is a short letter from Euphrasie, where she demonstrates the literary talent that we find again in the letters to Duguet, another Saint-Simonian and a close friend of hers.

Michel Chevalier's letter

Michel Chevalier was one of the most important Saint-Simonian theoreticians; he developed the religious, economic, and political views of the sect, as exposed for instance in his *Système de la Méditerranée*, published in 1832. He was a brilliant *polytechnicien* and was one of the early converts to Saint-Simonism and became very influential as the director of *Le Globe*, the most important Saint-Simonian newspaper. His quarrel with Enfantin during their imprisonment in *Sainte-Pélagie* from December 1832 until July 1833 is recorded in the *Archives of the Fonds Enfantin* at the Arsenal. From 1838 on he was a member of the *Conseil d'Etat*, of the *Conseil supérieur du Commerce*, and of the *Conseil supérieur de l'instruction publique*. Chevalier was technical counsellor of the famous *Compagnie des chemins de fer de l'Ouest* owned by two Saint-Simonians, the Pereire brothers, Olinde Rodrigues's cousins. From 1840 on he was professor of political economy at the *Collège de France*, and his successful career culminated at the beginning of the fifties, when he became a highly esteemed adviser of Napoleon III.

The letter by Michel Chevalier that we shall now discuss was written on 16 January 1832 and was addressed to Olinde Rodrigues in the midst of the conflict between the latter and Prosper Enfantin. He probably intended to demonstrate his unambiguous fidelity to the charismatic main ideas of Enfantin while, at the same time, not appearing disloyal to Olinde, to whom he shows a seemingly positive attitude. The coherent presentation of Enfantin's logic is for him more convincing and indeed more compelling than any evidence of abuse.

Seduction was Enfantin's customary way to defeat an adversary, his weapon to demonstrate the passivity and essential dependence of every woman and to win the contest with any male intellectual antagonist. He alone thought in a rational way, he believed, preferring ideas to emotions. For him, the others were slaves of a dark, irrational side. How painful to observe the naïvety of women and the blindness of men. Enfantin always remained an astonished spectator of the sentimental life

[9]After the schism Rodrigues refused to meet Bazard, even though the latter wanted to talk to him about private matters. FE*a*, f. 448. (See Notes at the end of the article for an explanation of FE*a*, FE*b*, and FAP.)

[10]Most of the relevant texts can be found in the *Bibliothèque de l'Arsenal*, in the *Bibliothèque Richelieu* (the former National Library), and in the *Bibliothèque Thiers*; more information may be obtained from articles written by Szajkowski (Szajkowski 1947) and Ratcliffe (Ratcliffe 1971, 1972, 1975). The three letters which are particularly important to us were copied together and are registered at the Arsenal in the *Fonds Enfantin, Archives* III, 7645, with the numbers 444 (Talabot), 445 (Chevalier), and 446 (Euphrasie).

around him,[11] and he found the parallelism of Bazard's plot and Rodrigues's affairs unquestionable.[12] Being an antagonist of both men, he planned and consistently developed the same strategy. How important was it for him to have the impulse to experiment with regard to the basic will to power? Was he striving after a vain ambition or did he believe in those experimental ideas that destroy every past human relation?

The situation may be analysed on the basis of Chevalier's letter. He describes the character of the protagonists and tries to present a general perspective and a solution. He begins with an analogy between the general and the particular, presenting them unified in the actions of the protagonists in the conflict for supremacy. Their actions are of a supreme importance, an 'immense affair,' more important than all other events inside the group. The inner life of the group, according to Chevalier, represents the beginning of a whole new world. Every action of a member, more so in the case of a charismatic chief, has a universal meaning. For Enfantin and for most of the Saint-Simonians, a private sphere does not exist: this view enables us to understand the nature of several conflicts among them.

Michel Chevalier subjects the most recent statements of Olinde Rodrigues to the rigorous logic of his chief Enfantin: '[y]ou told me that if *Père* Enfantin is wrong about Euphrasie, he is not the chief of the doctrine anymore. In this case you are the leader. But if he is right, you added, you are not Euphrasie's husband. It follows that the point [of the whole discussion] is in this sentence.' Chevalier expresses great esteem for Rodrigues, for his 'colossal' ideas and political projects. Yet he has misgivings about something else: the contemporaneous presence of three qualities in his friend's character. He is not sure if Olinde is priest, pontiff, and king at the same time, three indispensable requirements for the charismatic chief of the movement. We have to remember that four years of theoretical debates fixed the Saint-Simonian terminology and theory, but they never came to a definitive settlement or a peaceful agreement. Enfantin provoked all other members and tried to establish an open space for discussion, a sort of 'living theatre' that he called the 'living law.' In reality, all kinds of mixed feelings and ambitions made cooperation impossible. Because of this we cannot consider the Saint-Simonian ideas as a fixed corpus. Chevalier tried to describe what the Saint-Simonians needed, in particular, the qualities of the charismatic chief. This is a long, rich analysis in which we can isolate some contradictions, useful for us to understand the political lining up of ranks. He describes the past, the struggles within the movement when Bazard, Rodrigues, Enfantin, and others discussed and transcribed the development of Saint-Simon's ideas. In the years 1829 and 1830 they published the first volumes of the *Oeuvres de Saint-Simon et d'Enfantin*. Chevalier reminds Rodrigues how people found it difficult to deal with Bazard's 'impracticable rigidity' and how they did not wish to spend all their lives under 'Aristotle's baton-rod.' The difficulties continued with Rodrigues, and Chevalier's vocabulary becomes all the more important to us. He writes: 'as people keep away from Bazard, they keep away from you—Rodrigues—because they do not want to live among Aaron's guards,'[13]

[11]'I am, I confess, astonished at this blindness, as I had been at that of Bazard' (*Sainte-Pélagie*, 7 January 1833). It is one of Enfantin's personal remarks to a letter by Euphrasie Rodrigues, which he added during the months he spent in prison, after the trial in 1832.

[12]Ph. Régnier (Régnier 1989*b*, p. 169) analyses some Saint-Simonian conflicts, 'the fight for power within the group, the rivality to incarnate the "living law".'

[13]The allusion to Rodrigues's Jewish origin is quite clear.

and he adds, 'this is where I myself feel happy to exist.' Chevalier thinks that people want to live next to the 'king-pontiff,' the door to the *Sancta Sanctorum*. The 'sacerdotal attitude' that opens this door is not so much Bazard's strictness as Rodrigues's 'majestic boldness.' If Rodrigues had 'sublime ideas,' they are enclosed in the whole doctrine, written by Enfantin on 'Saint-Simon's mantle.' Every feature and gesture of Rodrigues, nevertheless, reveals a certain lack of harmony. Chevalier describes this quasimilitary attitude in a famous sentence—he actually compares him to a 'commander of a peaceful army of workers'—and stresses the word 'army,' asking his friend, whom he calls '*Père*,' to remember this word.[14] He tries to explain the difference between the people who prefer Rodrigues and the Enfantin followers. Everybody admires Rodrigues, many people love him, but some do not. Obedient, 'regular' people can appreciate him; the artists and the 'irregulars,' overwhelmed with their unbalanced emotions, are choked by such 'male frankness.' This openness wounds the sensitive natures, the rudeness has a negative impact on their delicate feelings, and they find his laconic character 'dry.' Chevalier's opinion diverges on this point; he thinks his friend is 'precise.' On the other hand, he knows why Reynaud, Transon, Laurent, and Leroux keep away from him and why some vivacious, light, graceful women feel themselves constrained next to him.[15] One can read, however, numerous letters to the contrary from women in the Saint-Simonian newspaper *Le Globe*. Riot-Sarcey (Riot-Sarcey 1992, p. 19) remarks that they often expressed their sympathy for Olinde Rodrigues. His constant battle for women's civil and political rights was much appreciated and gratefully acknowledged by female readers and followers alike. Mélanie, one of Olinde's several sisters, most of whom were Saint-Simonians, stirred the beginnings of female Saint-Simonian activism through discussions and writings.

Chevalier continues his letter imploring his friend to follow his reasoning and to meditate. It had been alleged that Euphrasie had been intimate with Charles Henry[16] (whose life, Rodrigues declared, she wanted to save and actually succeeded in doing so). If this was the case, Rodrigues's belief that she was still the purest of all women was a full and complete admission that Enfantin's theories on free love were correct. However, by Rodrigues's logic, if the allegation was true, her purity had gone, in which case they would not be the ideal couple that could aspire to the direction of the Saint-Simonian group.

Enfantin's believers did not fear the radicalism of logic in favour of an increased devotion to the leader, obviously detrimental to relationships between equals. Rodrigues was confronted with a dilemma, the dilemma of breaking with a friend, a violent rupture with his whole past. Suddenly everything changed; it was a sharp, abrupt awakening for him. We know that Rodrigues's reaction was to reject definitively Enfantin's authority and that he was hostile to such a violation of the privacy

[14] 'They are annoyed by your male frankness, that they qualify as rudeness, and by your laconism that they find dry and which I find precise. You have the tone and the gestures of a commandant of a pacific army of workers, of the army, notice this word. Such tone and gestures appear irritating, brutal, to their excessively sharpened senses... overexcited in the absence of "the woman".' Letter by Michel Chevalier to Olinde Rodrigues, FE*a*, f. 445.

[15] He cites an example, the young Anaïs, probably Anaïs Cazeaux, E. Talabot's girlfriend.

[16] Charles Léopold Elie Henry was an architect that was influential in the formulation of the Saint-Simonian doctrine. An intimate friend of the Rodrigues-Pereire family, he lived on rue Helder, near Olinde's house on rue Montholon, in the present-day 9*ème arrondissement*. He was a supporter of Enfantin, but after Bazard's schism he assisted Rodrigues. However, after it was alleged that he had an afair with Euphrasie, his collaboration ended.

of his family relations. In 1829 and also one year later, the discussions about the introduction of the religious hierarchy, the terminology, and the institutionalization of the *Père/fils* relationship were complex and protracted. The interlocutors of the Saint-Simonian debates were made unsure by their own reservations. Enfantin was not the only one to believe in the need for such an authoritarian structure. The fundamental idea of the hierarchy of talents created an ambiguity between liberty of expression and need for authority. The fanatical cult of Enfantin's personality was slowly developing inside the group and coincided with the growing difficulties in relations with the rest of the world. It was the beginning of a weak defence against the imminent danger that could end it all.

Rodrigues used stubbornness against seduction, the best intellectual weapon of his adversary. We find in the archives innumerable pages on Enfantin's 'magnetic' charm.[17] In the microcosm of the Saint-Simonian group the search for the ideal couple imposed an absolute transparency. The tone of the subsequent lines from Chevalier's letter becomes more and more emphatic about this point. In Chevalier's mind Euphrasie is a model of grace and elegance—only two of her many qualities—which contrast with Olinde's character, with his brutality, which she had denounced the previous day. He transposes the example of the Rodrigues couple onto the level of a public discussion, the daily debates about the doctrine and the search for the messianic couple. Enfantin's private life, because he is celibate, cannot be examined in a similar way, since he refused every traditional tie. After Barrault, d'Eichthal, and Lambert, he theorized about the need for the female half of the messianic couple, his future partner.

Chevalier comments on Euphrasie's declarations regarding herself and her husband: she said that it was not possible to find two natures which were so incompatible, so far away from each other, as the two of them. She added that they are an exemplary model of progress, since they managed to stay together in spite of their diversity. Chevalier casts doubts on the efficacy or even the possibility of an 'accommodation' for two incompatible characters living together. Disharmonious unions carry signs of humiliation; they implicate the renunciation of any individual expression.

Chevalier is internally split: he feels that he is dividing 'two beings, one is his adoptive father and the other an excellent heart, irresistibly attracting everybody around her.' We simply note the constant presence of the Oedipus complex characterizing several conflicts inside the group. Chevalier's attitude is similar to Enfantin's personal feelings towards his antagonists. A latent 'father-complex' made the supporters of Enfantin's theory aggressive toward the others, who defended a more traditional view of social and sexual relationships. Chevalier himself, on Enfantin's side, feels guilty. His attack (what he calls 'stabs') against Rodrigues could be cruel. 'With every stab, his heart jumps.' But his thoughts, 'imprisoned, would risk to choke him.' More and more, his reflections about the conversation with his friend convince him that his doubts are justified. Rodrigues is even 'abusive,' in judging his 'ideas, concerning men and women's intimate relationships.' Chevalier's attitude is protective towards Euphrasie Rodrigues, the supposed 'victim' of her partner's 'brutality.' The same attitude characterises Gustave d'Eichthal, another of Enfantin's devoted followers, inside his family and later in the Saint-Simonian

[17]FAP, 24609 II, f. 440, letter by the Saint-Simonian De la Porte to E. Talabot, or 24610 II, f. 415, letter by Moïse Retournet to Enfantin.

group. One of the main Saint-Simonian goals was to stop the suffering of abused women and oppressed social classes, an analogy that was essential to them. The unsolved tensions with the group's female members render a unique interpretation impossible. Enfantin at the same time oppressed and liberated the women around him. Chevalier's attitude also reveals his ambivalence: he fights for Euphrasie but discredits her and her partner, affirming that they cannot be leaders, as they are not the ideal couple at all. Olinde Rodrigues was not the King-Pontiff; he cannot enter the Sancta-Sanctorum's holy door. Now we can read that neither is he the Priest. Chevalier patiently explains that he speaks like the 'son who sees that his father loses his way and then forces himself to talk to him.' In Chevalier's mind Rodrigues had been looking all his life for a paternal figure, which he found when he met Saint-Simon. After Saint-Simon's death he returned to his parental home, which he left only the day he found somebody else—Chevalier is probably thinking about Enfantin—deserving his esteem. He cannot imagine Rodrigues's life without this support, 'this love coming from above.'

He reminds *Père* Rodrigues that the Priest is someone inspiring an artist and says that there are no artists around Rodrigues. On the contrary, he continues, the 'irregular natures' keep themselves away from him! The *Père* Enfantin was his priest, just as he himself was the priest to whom Euphrasie 'confessed' what she had to say, before talking to anybody else. Rodrigues is sure he is the Pope, although Chevalier also doubts that Euphrasie could be the female Pope. In reality he finds her good-hearted, beautiful, with the 'purest type of grace.' But, he asks, is she a 'priestess,' 'female industrialist,' and 'sage' all at the same time? This threefold division of the sacerdotal role is one of the principal themes of Saint-Simonian dogma. It sums up in itself the whole social hierarchy and includes the three Saint-Simonian orders: the general or social priest, the priest of science and art, and the priest of industry. We find a similar division of the human hierarchy in Saint-Simon's works.

The Saint-Simonians tried to reduce to human parameters 'the endless potential of love and the essence of the mystery of trinity.' The idea of trinity is a constant axiom in the new Saint-Simonian theology, which speaks of a social and political trinity of artists, sages, and industrialists. A further trinity within the concept of the dominion of sympathy is added to the moral trinity: the mobile, the constant, and the calm. Scientific and industrial progress is worthless if man does not express what he has learned and what he feels. In other words 'the wise and the industrialist cannot be real men if they do not intensively cooperate with the priest.'

The Saint-Simonians' search for the 'female partner' was characterized by abstraction. The women of the group meant their liberty or their emancipation in several different ways, but in their concrete lives, their ambitions could hardly follow the mystical evolution that the chiefs were pretending to guide. Likewise, Euphrasie could not be expected to overcome her human limits. Chevalier remembers that the day before Rodrigues had acknowledged that she was a brilliant artist.[18] We might say by this that she would inspire an artist, but Chevalier seems to ignore this obvious detail. (He probably meant exclusively male artists, when he asked Rodrigues

[18] We have already mentioned that she was a good pianist. This is also corroborated in 'Notes sur la famille Rodrigues (branche d'Olinde)' in the private archive of Dr. Jacques Béjot in Paris: she continued playing the piano until late in life, and in 1864 she gave piano lessons to Marie Regnault and Edmée.

'if he had ever inspired an artist.') He simply means that Euphrasie cannot be priestess, industrialist, and sage at the same time. Moreover, in consequence, she cannot be the female messiah, partner of the great priest. Her ineptitude matches Rodrigues's inadequacy.

Chevalier's tone becomes increasingly dramatic. '*Entreouvrir les lèvres*' when talking about such a subject becomes apostasy. It is Rodrigues's irregular 'apostolic situation' that imposes on him 'such a cruel obligation.' Chevalier is deeply convinced that he himself has to open up and talk about some reflections that he had elaborated the day before. He feels himself obliged to his friend and to Saint-Simon who 'dwells' inside Rodrigues; it is because Olinde calls Chevalier 'my son' that he wants to be worthy of this name expressing his love and admiration. It would be evil for him to bury them in his heart.

In March 1832 Olinde Rodrigues answered Chevalier in three published letters, which show how irreversible the break between Enfantin and Rodrigues was and why Enfantin forbade all his adepts to have any contact with Rodrigues. Even his cousin Isaac Pereire was not allowed to have business with him any more. The second letter describes the conflicts over finances that resulted from the schism, and it represents an official challenge to the cult surrounding Enfantin and to his 'seductive and diplomatic ways.'

Edmond Talabot's letter

Edmond Talabot was a lawyer and a jurist who after the July Revolution (1830) was named *substitut du procureur du roi.* He became a member of the Saint-Simonian *Collège* supporting Enfantin and was very active within the Saint-Simonian community, being responsible for proselitizing. His elder brother Paulin, an engineer, also participated in the movement; he is more famous than Edmond, through his interests in business and industry. Paulin was a close friend of Michel Chevalier from their early childhood in Limoges. During their retreat to Ménilmontant in 1832, Edmond created the famous Saint-Simonian costume, a doublet, double buttoned on the back, which required reciprocal assistance of the users. He died on 17 July 1832 during the cholera epidemic in Paris, and his death is solemnly described in the publication *Mort de Talabot.*

Edmond Talabot's letter, the last one of the filed letters, is dated 4 February 1832 and is addressed to Rodrigues. His tone is stricter and more solemn than Chevalier's. In his opinion Rodrigues disowns his past. The discussion was mostly on the several influences on the Saint-Simonian doctrine. Some authors quote the article of *Le Globe*, written by Duveyrier on 12 January 1832, where he recognises Charles Fourier's influence on Saint-Simonian doctrine. He mentions the work of Just Muiron (Muiron 1824), defining it as a sort of popular version of the previous book by Fourier, released as *Théorie des quatres mouvements et des destinées générales.*

An important historian like Luis Blanc[19] maintained that the ideas about man and woman that Enfantin later developed cannot in any way be considered the heritage of their master's theories. In Talabot's letter we read exactly the contrary of Blanc's views, of which he was firmly convinced. At the beginning of his letter he sums up the current debate: Rodrigues denies Enfantin's supremacy, 'the supreme chief's paternity.' Moreover, Rodrigues disagrees with Enfantin's solution for the

[19](Blanc 1841, Vol. III, Ch. 3, p. 120).

'moral question' and consequently also with his opinion on Euphrasie. That simply means for Talabot that all Rodrigues's opinions are wrong.

First, these opinions represent a denial of his master's teaching. In fact Saint-Simon wrote a letter at the end of his first work published in 1802, *Lettres d'un habitant de Genève à ses contemporains*, where he characterizes future government. He affirms that the qualities that characterize those whom God calls to represent him on earth are not any 'insignificant virtues' anymore. 'Insignificant virtues' like 'chastity' and 'continence' cannot be related to the 'true virtues,' which will be the 'talents,' 'the higher degree of talent.' Talabot explains that Rodrigues's master meant that the highest degree of talent would be linked to the 'highest morality.' This new morality must not be related to the obsolete virtues of Christianity. He continues quoting Saint-Simon's last book, where he recommends these new virtues to the Saint-Simonian priest. The priest must not draw the believers' attention to abstract ideas; he must employ 'sensual ideas,' combining them in order to provide 'the highest degree of happiness' possible for the human race during its earthly existence. It is in this way that Christianity will be reconstituted as a general, universal, and unique religion.

Although Enfantin's aim is to work effectively on the improvement of humanity's earthly destiny, Rodrigues finds that his 'combination of sensual ideas' is immoral. He does not see how a 'strong anathema still weighs on physical, sensual relations,' the relations of man with the world and with woman, because of the 'selfish individualisation' that characterizes them. Rodrigues cannot see that humanity is not destined to shame, that its sensual life will be glorified, just like its industrialization, the very day that it will be recovered with a religious character that socializes it. Rodrigues's mistake is to assign an importance to the exclusiveness of family life that it cannot and it must not have. His tendency is to substitute the selfishness of a clan for individual selfishness. He gives to the maternal feeling an exclusively selfish implication. This feeling, more than any other, has to be universalized; it cannot be reduced to maternity or to an individual feeling. On the contrary, its importance lies in 'the immense power to drive the next generations to love the future they desire, in fact, the woman is the centre of the human generation.'

Rodrigues's second fatal error, Talabot continues, is his 'anarchical behaviour.' He disagrees with the 'strongest men of the doctrine,' Michel Chevalier, Bouffard, Stéphane Flachat, Isaac Pereire, and Charles Henry, whom Talabot considered his closest friends. This information helps us to understand how the group was divided at this time. We can add this description to the one in the letter by Chevalier. Talabot says that Lambert, Barrault, and he himself cannot react as d'Eichthal, Duveyrier, and Hoart do. There is apparently a deep difference between these two groups of people. Talabot believes that Rodrigues is calling the 'inept to judge the apt'; this is serious because he was supposed to represent the voice of the master, who came to found the 'hierarchy of capacities.' The only possible result of such an attitude will be to demoralize himself and everybody who imitates his example. The movement is just emerging from a critical situation, so that this further problem must be solved. Talabot demands from the 'supreme chief'—Enfantin—to send Rodrigues to talk to 'every chief of service,' since he is rightly alarmed about their general frame of mind. Enfantin is asked to be present at this meetings.

Talabot's style and his vocabulary are impressive, a mixture of military strictness, radicalism, and old-fashioned haughtiness. His fidelity to Enfantin seems to be extremely fanatical; the way he judges the members of the group shows us just how varied his evaluation of his comrades is, and it is invaluable to know their divergent opinions. Also, they were placed in his mind in a hierarchical order according to his personal esteem of each of them. When we compare his relation to Rodrigues with Chevalier's anxious concern, Talabot sounds inflexibly impersonal. It is difficult to collect useful details on the personal level. We find only one positive allusion to Rodrigues's strength, although in the last days, Talabot asserts, he (Rodrigues) showed a 'blindness' not worthy of his qualities.

Talabot's letter fixes some interesting starting points for further thorough studies of Saint-Simon's works in comparison with later Saint-Simonian publications. The question is to relate the master's ideas on women and civil and political life to the heated discussions on this topic inside the group, some years later, in particular the limits of women's liberty, the definition of their role. However, a main problem remains: what does the 'moral question' consist of, which Talabot takes so seriously? To take up Enfantin's point of view would be useful, illustrating the several exchanges he had with other Saint-Simonians. This we shall discuss after we deal with the last letter.

Coming back to our initial concern, we focus on Olinde Rodrigues's personality, where we seem to find both aspects: science and religion, the symbolic and the semiotic. But, regarding the matter in a more specific way, it becomes evident that with more than one Saint-Simonian couple, the inconsistency between the two pressures discussed marks the conflict between man and woman as is the case of the incompatibility between Olinde and Euphrasie, described by Chevalier with great malice and precision.

Euphrasie's letter

The last letter, the most important one, is one which Euphrasie addressed to the *Père* (Enfantin) and was dated 25 November 1831. Philippe Régnier edited a partial publication together with an astute commentary on some small parts of it.[20] It is an intimate dialogue with Enfantin: the conflict between the chief and this woman—a woman of the group and the wife of a particularly important member and antagonist. The intimacy is striking: Euphrasie is close to her master as nobody else may be. The absence of blind obedience, however, does not imply her consciousness of being right. She is promising silent respect, at least in the future, and she hides her disobedience with anger, fearing something is wrong with herself. The will to speak, to express herself through debate, highlights the interior dialogue of her sensitive nature rather than any claim of a political nature. A comparison with Claire Bazard would be useful to illustrate the discrepancy between inner freedom and political consciousness. Claire put her faith in intellectual work, in the development of the doctrine: the achievement of political rights is her clear aim. She was energetic and extremely ambitious, whereas Euphrasie did not leave a noticeable trace of her political activity.[21] To the contrary, she seemed to aspire to

[20](Régnier 1989*a*, p. 73): 'In this Euphrasie chose to deny, whereas Claire, after speaking, repented, withdrew to the official morality, and went to the side of the fallen *Père* to direct the attack against the immoral *Père*, before taking refuge in absolute silence.'

[21]Not to forget that Julie Fanfernot (or Fanfernaut), one of the most radical activists during the July Revolution, met Olinde Rodrigues in a Saint-Simonians assembly, following him after

a harmonious community, neglecting the aggressiveness of personal ambition. She was a wife and a mother, but she put her group experience and the experience of having a family on the same level, which is the source of her ambiguity. Apparently, she did not reveal her intimate world as a wife or a mother when she related to others. In her letters to Duguet,[22] through the allegory of nature, the exuberance of blooming trees and flowers and the murmur of a brook,[23] she describes how she craves her realization as a woman and praises the vital experience within the Saint-Simonian community. The description of a dancing party on rue Monsigny, a few days after she had given birth to her child, represents the joy of her social existence and her emancipated relation to motherhood. The first invitation to life after a physically exhausting phase came to her through the glowing lights of the woods[24] in a moment of solitude. Her renewal came with the voluptuousness of dancing all night. Enfantin noted on the side of the copied manuscript that she pretends to ignore the fact that she had given birth a few days before, speaking about the feverish activities on rue Monsigny! We can imagine Enfantin's attempts to understand Euphrasie. In his eyes, she embodies the female nature. Her obscurity or any secret she would hold could be dangerous, as it would prevent the spell of the 'transfer' that guarantees the exercise of his power. She contradicts everybody's desire to open their hearts and confess all their troubles to the adored *Père*. In this way, he found it difficult to have a transparent relation to her, a 'vertical' (master/servant) one. Since the practice of power needs individuation of the failures and guilts of others, he required some tangible proofs of her inclination to lie or, at least, of her intrinsic ambivalence, of the darkness of her talented style.[25]

Euphrasie's writing may be claimed for *écriture féminine*, whose impulse is to inscribe femininity in itself, as it was conceived by Hélène Cixous and Xavière Gauthier in the mid-1970s. Euphrasie's letters are permeated by otherness, such as excess, subversion, play, and ambiguity and by fluid figurative language, seductive

---

his 'fall' (*disgrâce*). She became Euphrasie's best friend, and the two women spent the last years of their lives together. The Jewish poet (and former secretary of Saint-Simon) Léon Halévy wrote the famous *Souvenirs de Saint-Simon* in which he reported these important details on the Rodrigues-Pereire family. See *La France Littéraire*, n° 1, Paris, 1832.

[22]Charles Duguet was a trustworthy follower of Enfantin, but we do not know anything about his life before 1829 when he joined the Saint-Simonians. He was a lawyer and an old college friend of the Saint-Simonian Charles Lambert. Sébatien Charléty discovered that in 1840 after his return from Egypt he worked for the Ministry of Public Works. He seems to have been a probably involuntary tool in Enfantin's intrigues. His restless personality and his tendency to ecstatic visions led him to become a Catholic in the sixties. The letters by Euphrasie to Duguet that we find in the *Archives* are three in number. The first is dated 16 May 1831, the second, 29 December 1831, and the third, February 1832. The last was written after Euphrasie's exit from the Saint-Simonian community. They are all recorded in FE*a*, ff. 438, 439, 440, 441.

[23]'What a delight to hear the sweet bird songs mixing harmoniously with the murmur of running water, and to see in between the yielding branches the golden lines that snake, divide and shine like myriads of diamonds.' Lettre from Euphrasie to Duguet, FE*a*, f. 439r.

[24]'What a pleasure to breathe in the pure balm air, under the trees... of which the leaves softly agitated allow the azure of the sky to escape from their brilliant breast, like a rain of silver sparks.' Letter from Euphrasie to Duguet, FE*a*, f. 439r.

[25]This allusion to a possible connection between femaleness and style reminds us of a text, *Éperons*, by Jacques Derrida (Derrida 1976). We have to note that *Éperons* alludes, in another form, to the text of an exposé, also on Nietzsche's style (Derrida 1973). Claude Lévesque commented on the relation between '*la femme*' and '*le style*' in *L'innommable singularité de l'autre. Une lecture d'Éperons et d'autres textes de Jacques Derrida*, second chapter of his book (Lévesque 2002).

and challenging practices. Her vitality is not held back by 'economic considerations' and imposes itself as necessary. It is not by chance that critics focus their reservations on the supposed mystification of the *écriture féminine* and its lack of political purchase, because of the texts' insistence on the body and of its celebration of motherhood, a powerful source of women's oppression according to some feminist authors. All these elements, in fact, exist in Euphrasie's writing, as well as her refusal to define or theorize (also one of the most important points in Cixous's 'subversive' writing) and reflects her playful and strategic tendency to avoid submission to rules based in masculine theories (Cixous 1981).

Euphrasie's letter manifests a violent female protest against the supreme chief: her rancour rushes into words. Enfantin cannot find a female intellectual supporter at the end of his climb to power. After the Saint-Simonian experience the other women, his female antagonists, mostly survive between shame and lamentation, feeling themselves deeply guilty. Euphrasie's revenge expresses the particular nuance of her personal taste. Avenging herself, she does not renounce the appearance of lightness. The style of her correspondence as a whole is impressive; she handles words with a detached elegance, revealing and hiding in turn. She can play subtle games with Duguet by introducing cheerful monologues[26], and she can display humour[27] that mitigates any impression of animosity or of the lack of mirth that drives her into despair. She fears she may become boring and temperamental; the negative side of her sentimental nature is an extravagant melancholy. The excess of her dramatic power[28] usually follows disillusion or a breach of friendship after some trifling misunderstanding or sentimental trouble. Nature is her usual refuge after disappointment or stress. Euphrasie was probably losing influence and allies at this time, but she reacts against the passive acceptance of an image of herself, which others present to her, forcing this picture to materialize in the complex theoretical web of the Saint-Simonian debates. One of her last letters to Duguet lets her anger explode against every insinuation in a volley of protest.[29]

It is the same tense atmosphere that we perceive in the letter of 25 November, where she firmly refutes the accusation of being unfaithful, refusing any form of confession. The male jury surrounding Enfantin was not allowed into any sphere of her being, as arguably no one else was. Euphrasie is the only Saint-Simonian who does not conceive the accessibility of her own personality or of her intimate sphere as possible or even desirable. Moreover, she refuses to countenance this either in practice or in theory. The interesting point is that she does not consider this refusal

---

[26]'Here, if I know myself as a disputant, something to make you pay an honorable fine and reparation, unless you had already prepared a very excellent defence, which my good will shall listen to with its usual clemence. What, you do not say one word? You do not dare? That is pleasant; good, I shall respond for you.' Letter from Euphrasie to Duguet, FE*a*, f. 439r.

[27]'You do not like me, do you? If you were cross you would be very much wrong. Her I am very wise, very good, very gentle, very amiable; yes, I am all that here; I have slept 12 hours, make me a compliment, it is the first time in six months that I sleep so well.' Letter from Euphrasie to Duguet, FE*a*, f. 439r.

[28]'Duguet, human nature is very weak, I must agree, especially impressionable and I am the proof of this; your last invitation has done me harm, very much harm, [...] ah, tell me, I beg you, what evil spirit has inspired your words, your words that as a poison run into my heart and into my veins [...].' Letter from Euphrasie to Duguet, FE*a*, f. 441.

[29]'In my view you have behaved not as a friend, but as an enemy, you have, I would almost dare say tempted me, or, if you have not done it, you have very much judged me wrongly, in these two cases, since I am wise, because I know myself, you have mistaken me, I must therefore believe that I have been mistaken about you.' Letter from Euphrasie to Duguet, FE*a*, f. 441r.

in a conventional way: it does not represent for her a return to the traditional moral world. Here begins her inner process of differentiation, her detachment from the others, even from the intellectual Claire Bazard. The most striking passage of the letter is a sentence where she claims her right to refuse the chief's authority. She does not want to hide the negative implications of her difficult position; she 'is not sufficiently Christian at all' to prevent herself from feeling this unbearable hatred. The rejection of religious tradition is unquestionable. The metamorphosis the reader witnesses does not find support in a social acknowledgement. If it is true that Saint-Simonian correspondence was regarded as public within the community, it does not mean that debates or real acknowledgement followed automatically every expressed opinion.[30]

Euphrasie's letters represent a form of protest that is suspended between two worlds, imprisoned in her silence. She produces her own reconstruction of reality and her version of the facts that Enfantin considers a trap. On 7 January 1833, in the prison of *Sainte-Pélagie*, he noted on the side of her letter that 'she expected him to show the letter to Olinde,' whereas she implored him to abstain from that. She is a clever talker, a dialectician, and excellent at defending herself. The echo of her words gets lost in space, and her partners—luxuriant nature, the stars in the sky, the elation of music—cannot answer. She seems to detest authority itself: her letter begins with a long attack against the power of truth and against power itself. The 'hour of truth' means unfortunately for her an obligation to dissemble her true sentiments, an obligation to lie.[31] She restrains herself, hiding the wound that 'indifferent people or even worse touch without taking care.' The connection between authority and truth is intentionally the first sign of her dissent. We are surprised by an alternative attitude, which fully differs from other female reactions. We notice how even the most radical female followers of Enfantin identify the supreme chief's acknowledgement of their public confession with the official beginning of their militancy in the movement. Here we observe a clear opposition to this group dynamic.

Enfantin interprets her lines as a confession, this being what he expects to be the usual female approach towards him. He feels attracted by the 'darkness' of the female nature. The unusual reaction of Euphrasie becomes in his mind the norm of an abstract 'eternal feminine.' The only possible confession in this case is the confession of having been unfaithful to her husband. The supreme chief wants to believe in his own theories exploring the female's 'nature': analysing and focussing on the characteristics of femininity entail a systematic destruction of the texture of lies sustaining traditional unions. He decides to show the letter to Rodrigues when he celebrates his happiness as Euphrasie's husband. The obscure motives for Euphrasie's hatred could be a reason to doubt her fidelity. Her letter may be a proof of guilt. It is not a question of blame, but the victory of the principle of infidelity and sexual freedom; that apparently is what Enfantin was eager to demonstrate.

---

[30]The brilliant analysis of Régnier (Régnier 1994, p. 95) of the public character of the Saint-Simonians' private writings is not concerned with this detail.

[31]'Since the hour of the era of truth has struck, that is, since you alone are the chief, I cannot express a true sentiment, and retain constantly the truth ready to escape in my glances and in my actions.' Letter from Euphrasie to the *Père*, FE*a*, f. 446.

We may recall on this issue a document conserved in the *Bibliothèque Nationale.* This letter[32] contains an appreciation of Euphrasie, one sentence summarizing a deep fondness. The author is Charles Henry, already mentioned as an architect involved in the Saint-Simonian movement and a good friend of the Rodrigues family. Several authors have commented on this letter, alleging that it may reveal an intimate relationship, as already discussed. There are exceptions, however, the most important being Euphrasie herself. She writes that Olinde is the 'only one who exists' for her.

When our female protagonist speaks about her obligation to hide real feelings and thoughts, she expresses her lack of confidence in Enfantin's capacity. She finds that the satisfaction of his own 'moral ambition'[33] underlines that Enfantin's vanity was mainly of an intellectual nature, and she does not believe that he can be defined as a 'priest' or religious in any way. The dramatic tension comes from her feeling of duty to support her husband in his daily confrontation with life and activities, giving a positive energy, sharing his choices and experiences. She feels too weak to sustain him, being conscious of the destructive potential of a denial in this delicate phase.[34] She cannot really attack the Saint-Simonian leading characters and their ideas without blemishing her partner or hurting him to death! A tissue of lies follows from this that causes a split of her personality, an intolerable anguish. Enfantin seizes on this as corroborating his views on the double nature of the female character. Euphrasie ironically comments on the power of attraction of her double ego, the result of dangerous conflicts. She asks the leader to draw the public's attention—most of all her husband's attention—somewhere else. 'She will prefer lying to causing him despair.'[35] This sentence makes us understand why it was so important for Enfantin to copy the letter into the Saint-Simonian *Archives*, since he probably considered it as a proof of guilt.

In actual fact Euphrasie needs silence, being disgusted by the latest events.[36] She promises a blind obedience, a total abnegation; but this will be a submission without feelings, she will 'follow the star carrying her heart until it burns,' for the reason that Enfantin's words 'will be orders for her' too.[37] Euphrasie's letter

[32]FAP, 24610, ff. 12–14. Henry ends the letter to Charles Duveyrier writing that he is fond of 'Madame Rodrigues.'

[33]'I die if I shut up, I am not sufficiently Christian; I hate you for the harm you have done to me without any reason, and for the harm you have done to others in order to satisfy your ambition. Moral ambition, I mean. [...] No, I do not believe you are either religious or a Priest and if my poor heart were not so much used to suffering, I confess, I would hate you.' Letter from Euphrasie to the *Père*, FE*a*, f. 446.

[34]'I have no faith in you at all, and in front of Olinde I must appear happy and contented; I am weak and lack courage and my mission is to show strength and inspire others.' Letter from Euphrasie to the *Père*, FE*a*, f. 446.

[35]'I beg you not to pass on this letter to Olinde, he would not be able to endure it, and you would deprive him right now of his powers. In any case, I would be forced to lie, and rather than cause him dispair or his death, I would speak and act against my own convictions.' Letter from Euphrasie to the *Père*, FE*a*, f. 446r.

[36]She probably alludes to Claire Bazard and to the degeneration caused by the general anarchy. Margerin, seemingly exercising a nefarious influence, played the role of the seducer. Claire was supposed to be the 'victim of the moral disorder.' She was also an ambiguous companion of Enfantin, the two of them having a similar desire for power.

[37]'Do not expect from me anything other than submission, but affection never, I have now arrived at seeing an example of total duplicity, while you who desire to understand how far women can endure, would be satisfied.' Letter from Euphrasie to the *Père*, FE*a*, f. 446.

evinces the problematic approach to an authoritarian leader, even if he is despotic in the name of abstract freedom. She underlines the radicalism and the violence of such an experience, which transformed her expansive nature into an 'exemplary' split personality, developing two opposite attitudes in her mind.

The moral question

It was when the Saint-Simonian struggle raging around the moral question was nearing its climax that Enfantin's striving for what he understood as total transparency began. His proposal for a public debate to create new ideas and to wipe out what he considered the hypocrisy of traditional morality was intentionally radical.

The passionate Saint-Simonian discussion about the abuse of women and the urgent necessity for female emancipation began in 1828 and continued until 1833. When they tried to define women's role in society, the male Saint-Simonians surpassed 'the impulsive temperament' of their master. In spite of being officially inspired by 'New Christianity,' they mostly hearken to other sources. As Bulciolu writes, '[i]t is the reservoir of images in the liturgical Catholic life, to which especially Enfantin refers, which confines woman in her "natural" pattern and role': 'in this way she can become the heart of this social hierarchy.' Some of the Saint-Simonians were the first to doubt the radical impulse of this mysticism and to criticise Enfantin's totalitarian attitude. The most moderate attempts to limit his influence were expressed by Olinde Rodrigues, who intended to limit Enfantin's theorization of the supreme couple.[38] Rodrigues combined a more moderate thinking with drastic manners. His moderate views on the relations of men and women do not exclude radical claims concerning women's political and civil rights. Olinde's personal risks were all the greater: his whole family was libelled after the echo of the latest Saint-Simonian preaching had reached public opinion. Aggressive campaigns against the movement appeared in magazines and journals. The corrosive satire of some intellectuals such as Balzac deterred the majority of readers. People mocked or threatened Saint-Simonian followers in the streets.[39]

We can read in one of the rare descriptions of Rodrigues's personality that 'he was always ready to take power'[40] and that his authoritarian and quasimilitary attitude seemed unmannerly to people. We lack, however, reliable unbiased sources to understand the precise reasons, besides the obvious personal ones, that caused his dissent or his violent dispute with Enfantin, but we can recall his original intentions: to stress within the movement the importance of ethical behaviour and the urgent need of moral integrity. With his appeal of 1832 '*Aux Saint-Simoniens. Bases de la loi morale proposée à l'acceptation des femmes*' he differentiated his position from that of Enfantin after the schism in November 1831. He claimed Saint-Simon's heritage exclusively for himself and the paternity of the Saint-Simonian doctrine; furthermore he urged men and women to create a workers' union for pacifism to promote 'New Christendom.'

---

[38]He read his codification of the moral law on 17 October 1831, in front of Bazard, Enfantin, and the rest of the Saint-Simonian '*Collège*.'

[39]Enfantin tried deliberately to show in public this kind of aggression, exposing his opponents in Saint-Simonian periodicals and parades. The attacks against him and the movement became for him a sort of publicity. Cf. (Vidalenc 1960).

[40]Papiers personnels de Charles Lambert, *Bibliothèque de l'Arsenal, Fonds Enfantin*, 7803 and 7804, ff. 6-10.

Rodrigues did not exclude the dependence on variable factors that might affect the duration of a marriage. He accepted divorce only if it involved a new union, another engagement on a social level. Another couple of a 'superior degree' must approve the new couple, charged with a supreme task and power. The aim of every kind of union is scientific, industrial, and sacerdotal work and progress; married life must not be an actual 'institutional fiction' anymore. External unions and relationships are permitted, too; liberty cannot damage the exclusive passion between two spouses. That is why it is only through them that the bond among generations can be perpetuated. It will again be the supreme couple that reveals and establishes the limits of affections outside marital life.

In 1829, Olinde's much younger brother, Eugène Rodrigues, who was an important theoretician within the Saint-Simonian group, had explained why neither the female nor the male representatives, simple human beings, have to pretend to be united from generation to generation through the ages. Eugène was a member of the *Collège*, and the translation of Lessing's *Essai sur l'education du genre humain* (*Erziehung des Menschengeschlechts*, Berlin, 1780) and the *Lettres sur la religion et la politique* are his most important works. He had mystical inclinations and an introverted nature. Régnier commented that his earlier romantic renunciation of a lover and his religious convictions gave him an aura of sanctity. Isaac Péreire and he are supposed to be the models for Sigismond Busch in Zola's novel *L'Argent*.

According to Eugène Rodrigues, the supreme couple represents the whole of contemporary humanity, 'future and past humanity, belonging to God.' The great priest is 'the general man for excellence who represents the human species in a specific epoch with his acquired virtual existence.'[41] He is not the 'personification of the Living God's Son,' and his virtual union with the great priestess sums up the virtual existence of the whole of humanity. Bulciolu (Bulciolu 1980), as other authors have done, points out that the Saint-Simonian theory, under the appearance of an 'ardent spiritualism,' stayed in reality anchored to materialism. Their cult of the ego is also impressive, although they preached the development and the maturity of the whole of mankind. With women's collaboration, they were going to build 'the New Jerusalem which the Christians conceived and the creation of which God reserved to the Saint-Simonians.' The outsized Saint-Simonian ego unwittingly overlaps with the spirit of Christian doctrine: they simply imitated the idea of a Church founded on Christian revelation. Some of the Saint-Simonian documents postulate the unity of matter and spirit without considering 'the existence of another entity, the soul, that unites these two substances through history and leads them to the knowledge of God.' Bulciolu (Bulciolu 1980) does not pay attention, however, to the subtle ambivalence implying the glorification of the Christian tradition. The Saint-Simonian polemic is not openly revealed, but it represents a subtextual motive in their invectives. The antithesis between the past and the present existence of the 'religion of sorrow' and the future credo in 'joy and—free—love' is much more extreme than it seems at first sight. The eventual theoretical incongruity is a sign of the inconvenience caused by some Christian abstractions. The Saint-Simonian sanctification of 'pure' matter and of carnal pleasures does not

[41] Letter by E. Rodrigues to Ch. Duveyrier, 16 August 1829, FE*b*, ff. 436, 497, 498.

fit in with the Christian mediation of a spiritual presence.[42] The Christian concepts of soul and church do not satisfy the Saint-Simonian need for a new evaluation of the relation between matter and spirit. According to Christian theology, moreover, gender difference exclusively exists in regard to the flesh and the language of the flesh. On the contrary, the Saint-Simonians, ignoring the unique priesthood of Christ, describe gender difference on a physical and spiritual level.

We can also interpret the religious cult of the supreme chief as a derivation of a denatured Christendom, as d'Eichthal was to write, denouncing Enfantin's vanity, calling it the 'fatuity of Christ.' The unresolved conflict with religious tradition was expressed through indefatigable opposition as well as the hypertrophied development of the ritual and the theoretical elements of the tradition itself. Saint-Simon first conceived religion as a passionate form of relations between individuals. Enfantin added the symbolic *Père-fils* relationship to the new credo, combining religion and private-social affections. It was not by chance that it was Olinde Rodrigues, the only depositary of the master's oral tradition, who in an important speech approved the introduction of the second degree of initiation in the philosophical school. Since this transformation of the religious faith permeated the Saint-Simonian philosophy, the Saint-Simonians expected a revealed belief.

Eugène Rodrigues[43] tried to fix the limits of the Saint-Simonian 'great priest's divinity.' He asserted that the Saint-Simonian supreme chief has a limited power, that he is 'God's vicar only for a limited time and through the great priestess.' At the same time Eugène played a fundamental role in the institutionalization of the hierarchical parental relations, as well as in the intellectual evolution of the doctrine. It was all but impossible to limit Enfantin's self-deification, not only for theoretical reasons. In fact, the internal cohesion of the group was wholly based on his attractiveness and skill. The emotional life of the Saint-Simonians, with the libidinal relations that this implies, is related to a third entity, God. It would have been impossible to experiment so far without the assumption of a condescending superior. In fact their devotion to God moderates the danger of an unreasonable fixation on a unique person, the leader. The same devotion protected them from Enfantin's immoderate self-importance, the absolute power he had over others. Enfantin responsibly refused a full involvement with affective demands, discouraging passive expectations; on the other hand, his charismatic power was the only influence he could exercise in the last difficult phase of the Saint-Simonian conflicts. The antagonism he stirred up led to the ostracism of his male comrades, whereas the female intellectual partners, living a more complex experience, became estranged from common society and succumbed to the most ambivalent feelings. This is the case even with the radical Claire Demar: the women needed reassuring consideration by a male, paternal chief.

After their removal from the male hierarchy, women showed their capacity to react, but they nevertheless remained frustrated, as they were excluded from men's most important activities. In spite of their problematic cooperation in the several initiatives, it was from then on that they edited the most interesting publications and created the first political and female periodicals (Elhadad and Fraisse 1977).

---

[42]Saint-Simon's sentence illustrates this difference: 'The soul is the material point where all our strengths converge and where they diverge. Thought is a material attraction: it is one of the results of the movement of neural fluids' (Coignet 1883, p. 170).

[43]Courteault (Courteault 1925, p. 158) defines him as 'an illuminated young man.'

Enfantin, for his part, wrote several interesting pages on female rivalry and the irrational reluctance of women to acknowledge all female authority. His correspondence with Claire Bazard, Cécile Fournel, and other Saint-Simonians gives us an indication of this lack of confidence within the women's group.[44] The distrust of female leaders in Saint-Simonian women's circles is the equivalent of Enfantin's need for male collaborators. His studies at the *École Polytechnique* had led him to favour friendships with bankers and engineers. Enfantin could not leave aside the social prestige indispensable to create a political and religious movement, a pragmatic compromise that fitted his egotistic ambitions.

On one side, we can see Saint-Simonian ideological discourse, supposing that something is not expressed, as Valentin Pelosse (Pelosse 2001, p. 242) remarks, that in a society in a given period 'the dominant discourse offers resistance to every dangerous opposition,' a fairly obvious conclusion. It becomes impossible or difficult to talk about certain things; their formal exposition is also subjected to precise rules. We have to reconsider from this perspective the debate on women and its paradoxical evolution towards symbolization and myth. We should concentrate our attention more on the 'enounced/expressed' rather than on the 'omitted' parts of the Saint-Simonian discourse. It is maybe the only passage of Pelosse's work that could be criticized. The advantages of a psychoanalytical approach must not concentrate our attention on 'the missing parts' at the expense of the study of the available material.

A last important remark: there are several analyses, above all critical ones, on the significance of this particular form of symbolization, which characterizes and defines Saint-Simonian political and religious thought. Scholars have commented on the absence of a powerful female leader beside Enfantin, comparing this to the proliferation of symbols inspired by the myth of the Mother (Régnier 1992, pp. 31, 43). The Saint-Simonian ambivalence consists of this tendency to escape from everyday life, trying to enrich it through an 'enthusiastic presence.' Bulciolu (Bulciolu 1980) considers this attitude to be 'an immense pseudo-mystic hallucination', when considered within modern constraints. The Saint-Simonian debates, however, did entail some practical consequences, proletarian women appearing for the first time on the intellectual scene as heiresses and collaborators of the intellectual women with bourgeois or noble origins. Saint-Simonian speculations on the purely symbolic level showed its efficacy in the social field. This Saint-Simonian escapism was of transient value. During their new practical undertaking, too, after 1833,[45] they were confronted with the same crisis, revealing the myth itself, and its masked aims. Instincts were sublimated through this activity of symbolization that functioned as long as it was unconscious. The *École saint-simonienne* had long ceased to play the critical part that it had had before. But at the same time it opened several areas of thought, exploring possibilities in the field of inter-individual relationships

[44] FE*a*, ff. 40 to 50. The dialogue between Enfantin and Claire reveals all her difficulties in the women's group, which she directed during 1830 and after. Enfantin decided to leave the 'throne' beside him empty, declaring that women had to develop their debates and undertakings independently. Claire had some difficulties with her responsibility in the group's hierarchy; she felt too unsure or too weak to react to envy, competition, and critics.

[45] The Saint-Simonians, led by Enfantin, went to Egypt in 1833, searching for the mythical 'Mother.'

and favouring a new relation to the world and to the self, precisely through its ambiguity.[46]

In the Saint-Simonian world the symbolic function opposes life to death, without becoming 'a negative opium, masking the consciousness of the repellent figure of death'; as defined by Bergson (Bergson 1982, pp. 127, 137), on the contrary it represents a prospective dynamism, which is meant to improve the situation of man on earth. Symbols are therefore force fields, *Kraftfelder*, creating ties between the conscious and the unconscious, transcendence and immanence.[47] *Do* we have to reconsider several elements in the Saint-Simonian evasion from the licit, real world?

Conclusions

The long-standing lack of studies focussed especially on Rodrigues has led to simplistic interpretations of his personality, such as the current opinion that he had more puritanical, conservative views than other Saint-Simonians. On the contrary, he was an uncompromising supporter of social, political, and theoretical change and innovation. The importance he gave to the 'couple' as the most significant component of social life was immense; an attentive analysis of his writings on marriage has to be correlated to the acknowledgement of his constant concern to safeguard his privacy and family life. Most of all, we should not conceive each and every aspect of Rodrigues's multifaceted nature as singular, having no relation with one another. Purity (and not puritanism!) of geometrical forms or the precise rules describing physical attraction of spheroids exercised on him the same charm as metric or scores in poetry and music. His noteworthy knowledge of banking and his competence in economic questions were also deeply connected to his pursuit of innovating public life, introducing a more open-minded, unprejudiced vision of circulation and exchange. Far from being conservative or obsolete, he opened his life-long partnership with his wife Euphrasie to social and religious group-experiences, as they both were united in a common hope for emancipation and renewal of social life. Rodrigues could also be extremely encouraging and sensitive towards women's claims, helping some of the Saint-Simonian female activists to survive, supporting their political and social fight for equal rights.

All his life he sought harmony, harmony which he believed to be achievable through rigorous application of justice and more equitable laws. Although an idealist, Rodrigues was sufficiently conscious of the needs of real life; having a wide religious and cultural background, he already found the mere idea of adoring a man as God revolting. His infinite admiration for Saint-Simon and the high esteem he had for Enfantin nourished his conviction that his master Saint-Simon was an inspired prophet and the new leader Enfantin extraordinarily gifted, but he never believed them to be gods, as most Saint-Simonian male and female adepts did. Rodrigues's modesty and his talent for lengthy and private research enabled him to share and develop his ideas throughout a dialogue with an intellectual partner. Such was his relationship to Saint-Simon and later on to Enfantin, at least until the end of 1831.

---

[46]We find in Monique Rouillé's introduction to *Une correspondence saint-simonienne* an apology of Saint-Simonians' 'ambiguous attitude' creating illusions (the myth of a female messiah) but also real experiences. See (Louis and Rouillé 1990, p. 19).

[47](Freud 1961, 1968), (Gadamer 1960), (Lauf 1983).

The excesses of Enfantin's extreme narcissistic personality are unimaginable in the case of Rodrigues, whose character was far more austere; he preferred to solve his personal conflicts and his life-long experience of discrimination through a nonviolent steady fight for justice. For that reason, perhaps, he was less disposed than others to yield to seduction by idolized, charismatic persons. Some people describe him as an introverted nature: he seemed to refuse any dialogue, but at the same time he could be verbally aggressive. His drastic reactions often seemed harsh to persons whose aims were less clear than his and whose need for protection or authority surmounted their need for ethical rigour or thirst of justice.

Through an analysis of the two letters of Chevalier and Talabot, both followers of Enfantin, we have gained an insight into the nature of the tensions within the Saint-Simonian community. The whole problem, of course, hinges on the position of women, which was somewhat ambiguous if not contradictory for those in the Enfantin camp. On the one hand, as Rodrigues himself claimed, they accepted the need to recognize the equality of the sexes. On the other, because of the reverence due by the members of the sect to *Père* Enfantin and because of the strong personality and charisma of the latter, the women of the congregation (as well as the men) were in danger of being dominated by the high priest. Accusing Rodrigues of having conceived an 'exclusively male authority,' Enfantin seemed not to realize that his own exercise of power transformed itself more and more into a male autocracy, isolated from the outside world. No better way of demonstrating this dilemma may be found than through the study of Euphrasie's correspondence discussed here. With her 'strategy of silence' Euphrasie tried to break Enfantin's domination and his exploitation of women's need to be listened to in their wish to play a more active role in the community and in society. It is largely on this dilemma that the Saint-Simonian movement foundered, despite Olinde and Euphrasie Rodrigues's efforts to keep the sect along viable lines.

Acknowledgements

The author is most grateful to Professor Barrie Ratcliffe for useful discussions and for a generous critical reading of the manuscript which resulted in many improvements. I am most grateful to Mme. Stephane Toublanc, of the *Cercle de Généalogie Juive*, Paris, for much help, in particular for putting me in touch with Dr. Jacques Béjot, who most kindly made available to me his extensive unpublished notes on the Pereire and Rodrigues families.

Notes

The various files quoted are as follows, listed with the abbreviations used in these notes:

FE*a* *Bibliothèque de l'Arsenal, Archives*, Vol. III, *Fonds Enfantin*, Ms 7645.
FE*b* *Bibliothèque de l'Arsenal, Archives*, Vol. I, *Fonds Enfantin*, Ms 7643.
FAP *Bibliothèque Nationale, Fonds A. Péreire.*

Bibliography

Bergson, H. (1982). Les deux sources de la morale et de la religion (217th ed.). Presses Universitaires de France, Paris.

Blanc, L. (1841). Histoire de dix ans, 1830–1840, Vol. III. Pagnerre, Paris.

Bouglé, C. (1918). Le féminisme saint-simonien. In *Chez les prophètes socialistes*, pp. 55–111. F. Alcan, Paris.

Bulciolu, M. T. (1980). L'école saint-simonienne et la femme. Notes et documents pour une histoire du rôle de la femme dans la société Saint-Simonienne. 1823–1833. In *Études sur l'égalité*, (Corrado Rossom, ed.), pp. 7–42. Goliardica, Pisa.

Cixous, H. (1981). The laugh of the Medusa. In *New French Feminism* (Elaine Marks and Isabelle de Courtivron, eds.). Harvester, Brighton, pp. 245–264.

Cixous, H. and C. Clément (1986). The newly born woman, transl. B. Wing. University of Minnesota Press, Minneapolis.

Coignet, C. (1883). Saint-Simon et le saint-simonisme. In *La Nouvelle Revue: Cinquième Année*, **20**, 125–173.

Courteault, P. (1925). Un Bordelais saint-simonien. In *Revue Philomathique de Bordeaux et du Sud-ouest*, **28**, 152–166.

Derrida, J. (1973). La question du style in Nietzsche aujourd'hui. I. Intensités. In Coll. 10/18, pp. 253–299. Union Générales d'éditions, Paris.

Derrida, J. (1976). Éperons. Les styles de Nietzsche. Flammarion, Paris.

Durand, G. (1998). L'imagination symbolique. Presses Universitaires de France, Paris.

Elhadad, L. and G. Fraisse (1977). Femmes prénommées: les prolétaires Saint-simoniennes rédactrices de 'La Femme Libre' 1832–34, Ière partie and IIème partie, *Les Révoltes Logiques*, **4**, 24–62 and **5**, 8–29, Solin, Paris.

Freud, S. (1961). Vorlesungen zur Einführung in die Psychoanalyse (3rd ed.), *Gesammelte Werke*, Vol. 11, pp. 302–308. Fischer, Frankfurt am Main.

Freud, S. (1968). Die Traumdeutung. Über den Traum (4th ed.), *Gesammelte Werke*, Vols. 2 and 3, pp. 335–358. Fischer, Frankfurt am Main.

Gadamer, H. G. (1960). Wahrheit und Methode: Grundzüge einer philosophischen Hermeneutik. Mohr, Tübingen.

Kristeva, J. (1985) Au commencement était l'amour. Psychanalyse et foi, Paris A. Michel.

Kristeva, J. (1986) The Kristeva reader (T. Moi, ed.), Oxford University Press.

Lauf, I. D. (1984). Das Symbol in seiner Wirkung. Phänomenen der Transzendenz in Symbolforschung. In *Akten des ersten Symposions der Gesellschaft für Symbolforschung*, pp. 37–49. P. Lang, Bern.

Lévesque, C. (2002). Par-delà le masculin et le féminin, Aubier, Paris.

Louis, B. and M. Rouillé (1990). Une correspondance saint-simonienne: Angélique Arnaud et Caroline Simon (1833–1838). Côté-Femmes, Paris.

Manuel, F. E. (1956). The new world of Henri Saint-Simon. Harvard University Press, Cambridge, MA.

Muiron, J. (1824). Sur les vices des procédés industriels. Bossange Père, Paris.

Pelosse, V. (2001). Claire Demar. Appel au peuple sur l'affranchissement de la femme. Aux origines de la pensée féministe. A. Michel, Paris.

Ratcliffe, B. M. (1971). Les Péreire et le saint-simonisme. *Économies et Sociétés*, **5**, 1215–1255.

Ratcliffe, B. M. (1972). Some Jewish problems in the early careers of Emile and Isaac Pereire. *Jewish Social Studies*, **34**, 189–206.

Ratcliffe, B. M. (1975). Crisis and identity: Gustave d'Eichthal and Judaïsm in the emancipation period. *Jewish Social Studies*, **37**, 122–140.

Régnier, P. (1989*a*). De l'androgenèse du féminisme: les saint-simoniens, *Cahiers du CEDREF*, N. 1, pp. 65–79. Université Paris 7. Denis Diderot, Paris.

Régnier, P. (1989*b*). Le messianisme des saint-simoniens. *In Révolutions françaises et romantismes européens*, pp. 155–171. Istituto Italiano per gli Studi Filosofici, Naples.

Régnier, P. (1992). Le livre nouveau des saint-simoniens (ed. and comments by P. Régnier). Du Lérot, Tusson Charente.

Régnier, P. (1994). Usages saint-simoniens de l'épistolaire. In *La lettre à la croisée de l'individuel et du social* (M. Bossis, ed.), pp. 91–97. Kimé, Paris.

Riot-Sarcey, M. (1992). De la liberté des femmes: Lettres des dames au 'Globe', 1832–1832, Côté-Femmes, Paris

Rodrigues, O. (1832). Religion saint-simonienne, Olinde Rodrigues à Michel Chevalier, rédacteur du Globe. Lachevardière, Paris.

Vidalenc, J. (1960). Les techniques de la propagande saint-simonienne à la fin de 1831, *Archives de Sociologie des Religions*, No. 2. Centre National de la Recherche Scientifique, Paris.

Szajkowsky, Z. (1947). The Jewish Saint-Simonians and the socialist anti-semites in France. *Jewish Social Studies*, **9**, 33–60.

CHAPTER 5

Rodrigues's Early Work in Mathematics, 1813–1816

I. Grattan-Guinness

The subject of this paper is the theses with which Olinde Rodrigues obtained his doctorate in 1815 and some other papers that he published at the same time. First, we look at the system of higher education available to him. It fell into two distinct parts; he studied within one but published largely within the other.

The range of institutions for science and education

The first and superior part of the educational institutions was the system of engineering schools centred upon the *École Polytechnique*, which had been founded in 1794 as a new institution (Fourcy 1828). In the definitive arrangement from around 1800 it gave a basic training for two years, after which graduates spent an additional three or four years at an *école d'application* in more specialised study. After that the normal career lay in civil engineering or mining or as an officer in the army or navy; but some of the brighter young men went back into these institutions as teachers and eventually professors and/or graduation examiners ((Grattan-Guinness 1990, especially Chapter 2) and (Grattan-Guinness 2005)).

The creation of the *École Polytechnique* was part of a rapid partial repair of the higher education system, which had been suppressed in 1793. Another quick reformation was the '*Collège Royal*' in Paris, which was renamed '*Collège de France*' in 1794 and continued its special role of delivering high-level lectures to anyone who wished to attend without registration or examination. However, the mathematics teaching there was atypically elementary for several decades (Sédillot 1870).

The replacement of the score of universities around the country took much longer to effect, and then a different system was imposed. Not until 1808 did Napoléon sign a decree allowing the formation of the '*Université Impériale de France*' (Aulard 1911). The name was a misnomer, for no university in the usual sense of the word was formed: this institution was responsible for the administration of schools in the Empire, together with higher-level teaching in '*Facultés*' to replace the old universities. The Empire was divided into '*académies*,' again a quite nonstandard use of the word, for these were administrative regions; Paris was one, and each of the others was usually three or four contiguous *départements* with headquarters installed in the capital of one of them (for example, the *académie* of Montpellier). The traditional disciplines of law, theology, and medicine were joined by the disciplines of 'letters' (humanities) and of sciences. *Facultés* in these five discipline areas were set up in some *académies*; in Paris, and only there, all five were established. In addition, an '*École Normale*' was formed in Paris as the élite institution of the *Université*, with its own *Facultés* of letters and of sciences.

Successful students were given a *licence* as their first degree. For postgraduate work a student had to submit two theses and defend them before juries of professors. Supervision seems to have been nonexistent or at most on an informal basis. Success in a science *Faculté* led to the award of a *docteur ès-science* degree. Publication of the theses does not seem to have been compulsory, though (versions of) some of them were printed as short books or, as we shall see with Rodrigues, in the journal literature.

The work of the science *Facultés* overlapped with that of the engineering schools, but they were clearly inferior; for example, for a student it was much more prestigious to be selected for the *École Polytechnique* than for the *École Normale*, never mind attending at a *Faculté des Sciences*.

Various other professional institutions, governmental and private, further enriched the scientific scene of Paris. We need to note the *Société Philomatique de Paris*, a private but important scientific society that offered rapid publication of short papers (Berthelot 1888).

Rodrigues's unusual venue of publication

The life of Rodrigues is described in Chapter 2 in this volume. He did not gain admission to either of the leading Paris *écoles*: his Jewishness may have prevented his admission. Instead he took courses within the *Faculté des Sciences* in Paris and obtained his *licence*. The differences of reputation among the institutions were not of great import for his education in that, in the system of *cumul* then prevailing in France, academics (and others in public life) often held several posts at once. For example, Rodrigues's professor of physics was J. B. Biot (1774–1862), who also held the corresponding chair in the *Collège de France*. Biot wrote a large treatise on the subject in 1816, and at one point he credited Rodrigues and another student with a new formula concerning the refractive index of optical material (Biot 1816, p. 213).[1]

By the time of this honourable mention Rodrigues had successfully defended his *licence* and his doctoral theses. Most unusually, he published texts related to both awards and in a journal connected with the engineering part of the educational system. At its launch the *École Polytechnique* had set up a *Journal*, principally as a venue for publishing versions of lecture courses. While some courses did appear there, research articles were also published; by the 1800s they were dominating its pages, and teachers were publishing textbooks with publishers in Paris. So the professor of descriptive geometry at the *École Polytechnique*, J. N. P. Hachette (1769–1834), founded in 1804 the *Correspondance sur l'École Polytechnique*, a journal 'for the use of the students at the school.' It contained news about the school and mathematical problems and short articles, usually written by school students, recent graduates, or staff. Three volumes were produced up to 1816, when the school was modified in various ways by the Bourbon authorities when it was 'restored' after the removal of Napoléon; Hachette had been closely associated with

[1]Biot thanked 'Lévy et Rodriguez.' I take our hero to be the second student named here; no other candidate seems to be evident. The other student may have been the Belgian mathematician and mineralogist Armand Lévy (d. 1831), who was to write the calculus articles in the *Encyclopaedia metropolitana*.

Gaspard Monge (1746–1818) and so lost his *École Polytechnique* chair, and thus his journal ended.[2] However, as another example of the inferior status of the *Université*, Hachette kept his chair in the *Faculté des Sciences* of the Paris *académie*, which is presumably where he met Rodrigues.

This contact gave the young man his means of publication. Material related to the *licence* and the theses appeared in the third volume (1814–1816) of the *Correspondance*, and so did three other papers. While the subject matters of Rodrigues's contributions fitted in well with the mathematical range and level of the journal, the quantity, at around 60 pages, was an uncommonly large number, especially for an author who had no affiliation with the school.

Rodrigues's theses

Rodrigues defended his *licence* on 29 November 1813, during his twentieth year. He took a detail in differential geometry, a subject that he may have learned from Hachette. (For the background of this and the other papers discussed in this chapter, see also Chapter 2.) In Rodrigues (1814*b*) he gave a new formula for the angle of contingence c between 'consecutive tangents' on a curve in space, correct to the fourth power of the infinitesimal separation of tangents. From it the radius of curvature r was given by

$$(1) \qquad r = \mathrm{d}s/c = \mathrm{d}s/\sqrt{\{(\mathrm{d}\cdot \mathrm{d}x/\mathrm{d}s)^2 + (\mathrm{d}\cdot \mathrm{d}y/\mathrm{d}s)^2 + (\mathrm{d}\cdot \mathrm{d}z/\mathrm{d}s)^2\}},$$

where s was the arc length.

On 28 June 1815 Rodrigues defended his first doctoral thesis before a jury of the *Faculté* chaired by its dean, the textbook writer S. F. Lacroix (1777–1843).[3] Published as (Rodrigues 1816), it was a very good effort by comparison with others defended in mathematics at that time.[4]

Rodrigues had chosen 'the attraction of [homogeneous] spheroids,' an important topic in planetary mechanics especially due to the contributions of A. M. Legendre (1752–1833) and P. S. Laplace (1749–1827), and also Laplace's follower S. D. Poisson (1781–1840), professor of mechanics in the *Faculté*. Of the background literature he cited only Laplace's *Traité de mécanique céleste* (4 volumes, 1709–1805), presumably having (Laplace 1799) principally in mind. He may also have read some of the pertinent papers, which should have been readily available to him in Paris. On the geometrical side he may have been helped by a recent book on geometry by Monge's fervent student and follower Charles Dupin (1784–1873), another graduate of the *École Polytechnique* (Dupin 1813); as we shall see in the next section, he certainly knew of this book.

After forming Laplace's well-known equation for the external potential and Poisson's new equation for the interior potential, Rodrigues applied them to the external attraction to spheres, infinitely long cylinders, and ellipsoids. Two novelties attended this latter case.

Firstly, Rodrigues formed the potential as a triple integral over the ellipsoid, E, but instead of trying to integrate it over the coordinate variables, he decomposed E into infinitesimally thin confocal shells and integrated with respect to the radial

[2]The print run of the third volume was 500 copies. The three volumes were to be reprinted occasionally later.

[3]For the records of both doctoral examinations, see *Archives Nationales*, AJ[16] 5322, fol. 44 verso.

[4]For a detailed summary of it, see (Todhunter 1873, pp. 242–251, 275–277).

variable ((Rodrigues 1816, pp. 366–367); on quadrics in (Dupin 1813, see especially pp. 298–314)). He also adumbrated the usual analysis of attraction to a point P external to E by considering also an ellipsoid confocal to E and passing through P and found various results relating the ellipsoids (Rodrigues 1816, pp. 370–373). One of them was a theorem on the potentials of two confocal ellipsoids that had recently been proved by the Scottish mathematician James Ivory (1765–1842), who was then making several contributions to potential theory. The theorem had aroused considerable interest in Paris; in particular, as their editor for mathematical articles, Poisson had written an account for the *Société Philomatique* (Poisson 1812), which Rodrigues cited.

Secondly, for 'ellipsoids infinitely little different from a sphere' Rodrigues expanded the generating function of the Legendre functions (to use the modern names; see Chapter 6 for the definition of these functions). He explicitly noted that its convergence needed to be established (Rodrigues 1816, p. 375)—not then a routine concern, although he did not amplify his warning with any analysis.[5] He found the complete integral to the associated Legendre equation, with two constants of integration (Rodrigues 1816, pp. 376–378):

$$(2)\quad \mathrm{d}^{m+n}\{(1-\mu^2)^n[F+K\int(1-\mu^2)^{-m-1}\,\mathrm{d}\mu]\}/\mathrm{d}\mu^{m+n},\ \text{with } F \text{ and } K \text{ constants.}$$

From this form he then found the compact formulae now named after him for the associated functions,

$$(3)\qquad J(1-\mu^2)^{n/2}(\mathrm{d}^{m+n}(1-\mu^2)^m/\mathrm{d}^{m+n}\mu),\ \text{with } J \text{ a constant,}$$

and established their weighted orthogonality over a finite interval (Rodrigues 1816, pp. 376–382).

Rodrigues was working on a popular topic, so the question of independence arises. Series expansions related to (3) had been found recently by Legendre in a volume on elliptic and other integrals (Legendre 1815, pp. 263–275) and by Ivory in the paper (Ivory 1812), but Rodrigues claimed independence (Rodrigues 1816, p. 376).[6] Again, Rodrigues's formation of potentials and also his use of confocal ellipsoids resembled a recent exploration of 'the attraction of ellipsoids' made by C. F. Gauss (1777–1855) (Gauss 1813); but Rodrigues did not cite it, and apparently nobody mentioned it to him.

Poisson presumably accepted without qualms the thesis version of Rodrigues's analysis, but in 1837 he criticised Rodrigues's use of shells. The occasion was yet another dispute with Louis Poinsot (no friend of Poisson) (1777–1859) (a graduate of the founding enrollment of the *École Polytechnique*), in a suite of short exchanges in the *Comptes rendus* of the *Académie des Sciences*. Poinsot was happy with Rodrigues's procedure, but Poisson (Poisson 1837) appealed to his own recent treatment (Poisson 1835) of the attraction of ellipsoids, which he divided into *similar* ellipsoidal shells (that is, shells with the same eccentricity) whereas Rodrigues had used *confocal* shells. Surely either form of dissection of the ellipsoid is possible,

---

[5]In a later paper on the attraction of ellipsoids Poisson took the convergence of the series for granted (Poisson 1826, pp. 364–371). He cited Ivory's theorem in this paper (Poisson 1826, p. 358), but not Rodrigues's version.

[6]Conversely, presumably the reappearance of (3) in (Ivory 1824, p. 93) was uninfluenced by Rodrigues. (Todhunter 1873, p. 248) mistakenly located it in (Ivory 1822).

if the details are handled properly, such as the form of the inner limiting shell;[7] Cayley (Cayley 1858) was to rework Rodrigues's approach in a rather tidier manner but without significant reform. However, when Todhunter reviewed the 1837 dispute, he judged in favour of Poisson (Todhunter 1873, pp. 381–387).

Rodrigues defended his second thesis on 5 July 1815. His topic was described as the rotation of a heavy body of revolution, which sounds close to the short paper (Rodrigues 1814*a*) on the 'rotation of free bodies'; he did not mention any (forthcoming?) thesis there, but maybe it was a part of a draft or preliminary version. He started out from the large textbook on mechanics that Poisson had published a few years earlier from his teaching of the subject at the *École Polytechnique* (Poisson 1811, Vol. 2, pp. 130–149). Taking from the book the basic Euler equations for the rotation of a solid body involving its moments and products of inertia, Rodrigues manipulated them into an integral form expressed as functions of angles pertaining to the body. He then used the resulting integral to provide formulae relating the directions of various axes of rotation of the body. Years later he would find a far more striking theorem on rotation, as discussed in Chapters 2 and 8.

Rodrigues's other papers in the *Correspondance*

In another short note on mechanics Rodrigues (Rodrigues 1814*c*) took further inspiration from Poisson's textbook, this time concerning the motion of a mass-point along a given curve (Poisson 1811, Vol. 1, pp. 370–381). Using again techniques from differential geometry, Rodrigues showed that the resistance to the motion made by the curve opposed the combination of the force moving the point and the mathematical force involving the radius of curvature of the curve, which he called the 'centrifugal force.'

A third short piece on mechanics used another major text of the time, the first volume of the second edition of *Mécanique analytique* of J. L. Lagrange (1736–1813). He had made the principle of least action a main assumption underlying dynamics (Lagrange 1811, Part 2, Sect. 3); noting that the principle reduced to the optimisation of 'live forces' acquired by a mathematical system when changing from one situation to another, Rodrigues (Rodrigues 1815*b*) used Lagrange's own method of multipliers to incorporate the equation of the conservation of these forces as a constraining condition in nonholonomic situations. This was a nice finding, and in a posthumous edition of Lagrange's book Joseph Bertrand (1822–1900) added an enthusiastic editorial reference to the paper (Lagrange 1853 ed., p. 281); when a further edition was prepared as part of the edition of Lagrange's works, Gaston Darboux both retained Bertrand's note and added a further reference to the paper in another Bertrand note (Lagrange 1888, pp. 323, 318). Rodrigues's paper was also translated into German in an edition of a quartet of important texts on principles of mechanics (Jourdain 1908, pp. 12–15).

---

[7]Neither Rodrigues nor Cayley explicitly discussed the inner limiting shell. Parallel planar sections of a family of similar ellipsoids can be taken to be parallel sections of a cone, of which the limiting case is its vertex; thus this shell has to enclose the centre of the ellipsoid, which corresponds to this vertex. For a family of confocal ellipsoids the limiting case is the interior and the boundary of the common focal ellipse E, which lies in the common section of the ellipsoids defined by their major and mean axes; so the inner limiting shell has to include E. For a review of the history of focal curves of quadrics from Dupin onward, see (Staude 1904, esp. Chapter 7); the basic properties are nicely presented, with excellent diagrams, in (Hilbert and Cohn-Vossen 1952, pp. 19–25).

In the final paper to be considered, Rodrigues (Rodrigues 1815*c*) treated differential geometry; it seems to be a development of the matter handled in the *licence* note. This time the inspiration came from the fourth edition of Monge's textbook on 'analysis applied to geometry' (Monge 1809) and from the book (Dupin 1813) on geometry mentioned earlier (cited in (Rodrigues 1815*c*, p. 164)); he might also have read the important paper (Dupin 1807) on the principal radii of curvature of a surface. Rodrigues ran through the basic theory of lines of curvature on a surface using a coordinate system defined from the direction cosines of those lines relative to the (arbitrarily) given Cartesian system (Rodrigues 1815*c*, pp. 162–167, including an explanatory footnote by Hachette). Finding some seemingly new formulae involving the radii of curvature of the surface, he applied them to various specific surfaces (Rodrigues 1815*c*, pp. 166–173).

The rest of the paper was based upon a theorem about a double integral, in which he noted some stimulation from Poisson (Rodrigues 1815*c*, p. 177). A short summary of it appeared with the *Société Philomatique*, initialled by Poisson; since it refers always to 'Rodrigue,' he seems to have written it himself, but I cite it as (Rodrigues 1815*a*). Take z as a function of x and y and use the usual Euler letters 'p, q, r, s, t' for the first- and second-order partial differential coefficients. By considering the formula involved in the change of independent variables, Rodrigues showed that

$$(4) \qquad \iint U(p,q)(rt - s^2)\,\mathrm{d}x\,\mathrm{d}y = \iint U(p,q)\,\mathrm{d}p\,\mathrm{d}q \text{ for 'any' function } U,$$

so that the value of either integral depended only upon the limits of integeration and not on the surface involved. He also showed that the first variation of $(4)_1$ was zero, as should be under these circumstances (Rodrigues 1815*c*, pp. 173–176).[8]

Both texts also contained the special case given by

$$(5) \qquad U \doteq (1 + p^2 + q^2)^{-3/2},$$

when $(4)_1$ became the double integral of the inverse product of the principal radii of curvature of a surface; by the previous result, it was equal to $(4)_2$, which applied to the sphere. Therefore the integral also gave the surface area of a specific sphere or, equivalently, a specific portion of the surface area of the unit sphere, the portion determined in each case by the original limits of integration. This result was proved for Rodrigues by Jacques Binet (1786–1856) (Rodrigues 1815*c*, p. 177), another graduate of the *École Polytechnique* and a passionate Catholic who was about to be appointed Director of Studies there as another change made at the school.[9] In the rest of the paper Rodrigues evaluated integrals of the form of (4) for various other surfaces, such as the ellipsoid, and in a three-page '*Addition*' he furnished alternative proofs of his main results.

Concluding remark

These last contacts show that an outsider in the wrong religion could enjoy working relationships with more orthodox figures, such as the eminent Poisson,

[8] Writing this note may well have inspired Poisson to have his own paper placed soon afterwards with the *Société Philomatique de Paris*, where he considered the application of the calculus of variations to double integrals (Poisson 1816); however, Rodrigues was not mentioned.

[9] In his *cumul* Binet also assisted in the astronomy teaching at the *Collège de France* and later took it over.

a soon-to-be-marginalised Hachette and an emerging personage like Binet. Rodrigues's 60 pages in the *Correspondance* exhibited an excellent talent surely about to flower into a career as a talented mathematician. The time was optimal for continued work, for he was among a quite exceptional cohort of contemporaries and elders, thanks especially to the manner of selecting and encouraging students at the *École Polytechnique* and the *écoles d'application* and of securing professional employment afterwards. Let us note some main currents; the details are recorded in (Grattan-Guinness 1990, Chapters 8–16).

Laplace, Biot, and Poisson, along with several physicists and chemists, were developing an ambitious programme of mathematicised molecular mechanics and also physics. They were assisted especially by E. L. Malus and also D. F. J. Arago in their ballistic optics. Poisson himself had considered electrostatics already in 1812, and later he took up magnetism. However, the programme was met by formidable alternative theories that in some way or another were not based upon molecular modelling: heat diffusion with Joseph Fourier; wave theories of light by A. J. Fresnel, with some support from Arago (who changed sides) and A. M. Ampère, and later A. L. Cauchy; electrodynamics and electromagnetism by Ampère; energy/work mechanics in the hands of various engineers, especially C. Navier, G. G. Coriolis, J. V. Poncelet; and stress-strain elasticity theory from Cauchy and Navier, with Poisson in competition. Much more work in astronomy and potential theory (principally concerning equipotential surfaces) was done by Laplace and Poisson and by various others. In addition, Monge's legacy and Hachette's endeavours gave new life to geometry, not only descriptive or differential. Dupin produced other nice results and applications, Poncelet published his important pioneering book on projective geometry in 1822, and other significant figures include J. D. Gergonne, M. Chasles, C. J. Brianchon, and G. P. Dandelin. In the purer branches of mathematics Cauchy introduced his formulation of mathematical analysis in (controversial) teaching at the *École Polytechnique* and also laid the foundations of complex-variable analysis.

But Rodrigues was not to join in any of this activity: he did not produce (or at least did not publish) any more mathematical work for 20 years[10] and instead became deeply involved with some social movements, as described in Chapters 2 and 3 in this book. Maybe, and rightly or wrongly, he perceived his Jewishness as a block in the aggressively Catholic atmosphere, which was to hold from the Restoration in 1815 until the French people became once again disgruntled in 1830 and had another revolution. However, his failure to continue to work in mathematics is a great pity, given the propitious circumstances of the time for mathematics and his own short but intensive and productive start in his early twenties.

Bibliography

Aulard, F. V. A. (1911). *Napoléon et la monopole universitaire: origines et fonctionnement de l'Université Impériale.* Colin, Paris.

Berthelot, P. E. M. (1888). Sur les publications de la Société Philomatique et sur ses origines. *Journal des Savants*, pp. 477–483. *Also in Mémoires publiés par la Société Philomatique à l'occasion du centenaire de sa fondation.* Gauthier-Villars, Paris, pp. i–xvii.

[10] In a rare slip, when listing Rodrigues's contributions in the catalogue of scientific papers, the Royal Society placed the seven earlier items discussed here as written by 'Rodrigues, (*Dr.*)' and his later ones by the apparently different author 'Rodrigues, *Olinde*, (Royal Society 1871, p. 250).

Biot, J. B. (1816). *Traité de physique expérimentale et mathématique*, Vol. 3. Deterville, Paris.

Cayley, A. (1858). On Rodrigues's method for the attraction of ellipsoids. *Quarterly Journal of Mathematics*, **2**, 333–337. [Reprinted in *Collected mathematical papers*, Vol. 3, pp. 149–153.]

Dupin, P. C. F. (1807). Essai sur la description des lignes et des surfaces du second degré. *Journal de l' École Polytechnique*, Séries 1, **7**, Cahier 14, 45–83.

Dupin, P. C. F. (1813). *Développements de géométrie.* Courcier, Paris.

Fourcy, A. L. (1828). *Histoire de l'École Polytechnique.* École Polytechnique, Paris. [Reprinted Belin, Paris (1987).]

Gauss, C. F. 1813. Theoria attractionis corporum sphaeroidicorum ellipticorum homogeneorum methodo nova tractata. *Commentationis Societatis Segiae Scientiarum Gottingensis Recentiores*, **2** (1811–1813), cl. math., 24 pp. Also in *Monatliche Correspondenz*, **28**, 37–57, 125–134. [Reprinted in *Werke*, Vol. 5, pp. 1–22. See also the summary paper on pp. 289–296.]

Grattan-Guinness, I. (1990). *Convolutions in French mathematics, 1800-1840: From the calculus and mechanics to mathematical analysis and mathematical physics* (3 vols.), Birkhäuser, Basel and Deutscher Verlag der Wissenschaften, Berlin.

Grattan-Guinness, I. (2005). The Ecole Polytechnique, 1794–1850: differences over educational purpose and teaching practice. *American Mathematical Monthly*, **112**, 233–250.

Hilbert, D. and Cohn-Vossen, S. (1952). *Geometry and the imagination.* Chelsea, New York.

Ivory, J. (1812). On the grounds of the method which Laplace has given in the second chapter of the third book of his Mécanique céleste for computing the attractions of spheroids of every description. *Philosophical Transactions of the Royal Society, London*, **102**, 1–45.

Ivory, J. (1822). On the expansion in a series of the attraction of a spheroid. *Philosophical Transactions of the Royal Society, London*, **112**, 99–112.

Ivory, J. (1824). On the figure requisite to maintain the equilibrium of a homogeneous fluid mass that revolves upon an axis. *Philosophical Transactions of the Royal Society, London*, **114**, Pt. 1, 85–150.

Jourdain, P. E. B. (ed.) (1908). *Abhandlungen über die Prinzipien der Mechanik von Lagrange, Rodrigues, Jacobi und Gauss.* Engelmann, Ostwalds Klassiker, no. 167, Leipzig.

Lagrange, J. L. (1811). *Mécanique analytique* (2nd ed.), Vol. 1. Courcier, Paris. [Reprinted in (Lagrange 1888).]

Lagrange, J. L. (1853). *Mécanique analytique* (2nd ed.), Vol. 1, new ed. by J. Bertrand, Mallet-Bachelier, Paris. [Reprinted in (Lagrange 1888).]

Lagrange, J. L. (1888). *Oeuvres*, Vol. 11. Gauthier-Villars, Paris. [Based on (Lagrange 1853).]

Laplace, P. S. (1799). *Traité de mécanique céleste*, Vol. 2. Duprat, Paris. [Reprinted 1829 and as Vol. 2 of the *Oeuvres* (1843) and the *Oeuvres complètes* (1879).]

Legendre, A. M. (1815). *Exercices de calcul intégral sur divers ordres de nombres transcendantes et sur les quadratures*, Vol. 2. Courcier, Paris.

Monge, G. (1809). *Applications de l'analyse appliquée à la géométrie, à l'usage de l'École Polytechnique* (4th ed.). Bernard, Paris.

Poisson, S. D. (1811). *Traité de mécanique* (1st ed.) (2 vols.). Courcier, Paris.

Poisson, S. D. (1812). Mémoire sur l'attraction des ellipsoïdes homogènes; par M. Yvory [sic]. *Bulletin Scientifique de la Société Philomatique de Paris*, **3** (1812–1813), 176–180.

Poisson, S. D. (1816). Sur le calcul des variations relativement aux intégrales multiples. *Bulletin Scientifique de la Société Philomatique de Paris*, 82–86.

Poisson, S. D. (1826). Mémoire sur l'attraction des sphéroides. *Connaissance des temps*, (1829), 329–379.

Poisson, S. D. (1835). Mémoire sur l'attraction d'un ellipsoïde homogène. *Mémoires de l'Académie des Sciences*, **13**, 497–545.

Poisson, S. D. (1837). Addition aux remarques insérées dans le Compte rendu de la séance dernière. *Comptes rendus de l'Académie des Sciences*, **7**, 1–3.

Rodrigues, O. (1814*a*). Sur le mouvement de rotation des corps libres. *Correspondance sur l'École Imperiale Polytechnique*, **3** (1814–1816), 32–36.

Rodrigues, O. (1814*b*). De l'angle de contingence d'une courbe à double courbure. *Correspondance sur l'École Imperiale Polytechnique*, **3** (1814–1816), 36–37.

Rodrigues, O. (1814*c*). Sur la résistance qu'éprouve un point matériel assujetti à se mouvoir sur une courbe donnée. *Correspondance sur l'École Imperiale Polytechnique*, **3** (1814–1816), 37–39.

Rodrigues, O. (1815*a*). Sur quelques propriétés des intégrales doubles et des rayons de courbure des surfaces. *Bulletin Scientifique de la Société Philomatique de Paris*, 34–36. [Signed 'P.' by Poisson.]

Rodrigues, O. (1815*b*). De la manière d'employer le principe de la moindre action, pour obtenir les équations du mouvement rapportées aux variables indépendantes. *Correspondance sur l'École Imperiale Polytechnique*, **3** (1814–1816), 159–162.

Rodrigues, O. (1815*c*). Recherches analytiques sur la théorie des lignes et des rayons de courbure des surfaces, et sur la transformation d'une classe des intégrales doubles, qui ont un rapport direct avec les formules de cette théorie. *Correspondance sur l'École Imperiale Polytechnique*, **3** (1814–1816), 162–182.

Rodrigues, O. (1816). Mémoire sur l'attraction des sphéroids. *Correspondance sur l'École Imperiale Polytechnique*, **3** (1814–1816), 361–385.

Royal Society (1871). *Catalogue of scientific papers (1800–1863)*, Vol. 5. Royal Society, London. [Reprinted 1968, Scarecrow, New York.]

Sédillot, L. P. E. A. (1870). Les professeurs de mathématiques et de physique générale au Collège de France, Part 4. *Bullettino di Bibliografia e di Storia delle Scienze Matematiche e Fisiche*, **3**, 107–170.

Staude, O. (1904). Flächen 2. Ordnung und ihre Syteme und Durchdringungskurven. In *Encyklopädie der mathematischen Wissenschaften*, Vol. 3, Pt. 2, 161–256 (article IIIC2).

Todhunter, I. (1873). *A history of the mathematical theories of attraction and figure of the earth from the time of Newton to the time of Laplace*, Vol. 2. Macmillan, London. [Reprinted 1962, Dover, New York.]

CHAPTER 6

The 1839 Paper on Permutations: Its Relation to the Rodrigues Formula and Further Developments

Richard Askey

Introduction

Two seemingly unrelated results of Rodrigues are the focus of this chapter. The first was included in his dissertation and was published in (Rodrigues 1816). To make it easier for the reader, we will use current notation.

Legendre polynomials, $P_n(x)$, are polynomials orthogonal on $(-1, 1)$.

$$\begin{aligned} \int_{-1}^{1} P_n(x)P_m(x)dx &= 0, \qquad m \neq n, \\ &= \frac{2}{2n+1}, \qquad m = n. \end{aligned} \tag{1}$$

One of the results Rodrigues found is

$$P_n(x) = \frac{1}{2^n n!}\frac{d^n}{dx^n}(x^2-1)^n. \tag{2}$$

This formula was rediscovered by Ivory (Ivory 1824) and by Jacobi (Jacobi 1827) and was known as the Ivory and Jacobi formula for many decades. (Ivory (Ivory 1812) arrived at an expression from which formula (2) can be derived, but he did not do this until his paper of 1824; so the 1812 paper cannot claim priority for Ivory over Rodrigues.) Heine (Heine 1861, pp. 10–11) gives (2) and references the papers of Ivory and Jacobi. There is a joint paper of Ivory and Jacobi (Ivory and Jacobi 1837) which proves (2). They published this in Paris because (2) was an important formula that was not known in France.

It seems that the first reference to (Rodrigues 1816) was (Cayley 1858). Julio González Cabillon kindly drew my attention to this paper. However (2) was not included since Cayley dealt with other parts of Rodrigues's work. The first reference to the occurrence of (2) in (Rodrigues 1816) is (Hermite 1865*b*, p. 346). Heine (Heine 1878) included a reference to this work of Rodrigues, and it was this book which led to (2) and extensions of it being called Rodrigues's formula or formulas of Rodrigues type.

The other result of Rodrigues seems to be completely unrelated. The article (Rodrigues 1839) contained a solution to a problem of Stern (Stern 1838) which had earlier been solved by Terquem (Terquem 1838). However, Rodrigues extended the problem to something more important. What he did was to find a generating function for the $n!$ permutations of $\{1, 2, \ldots, n\}$ where the coefficients in the generating function are the number of permutations of $\{1, 2, \ldots, n\}$ with k inversions.

This result was found long before others were interested in this problem. MacMahon (MacMahon 1916) rediscovered Rodrigues's generating function and extended it in important ways. However, the first reference to (Rodrigues 1839) seems to be (Carlitz 1970). Carlitz wrote that it was one of his students, Charles Church, who rediscovered Rodrigues's paper. We will return to these combinatorial results later and eventually tie this up with polynomials which have Rodrigues type formulae.

The work on Legendre polynomials

Legendre polynomials arose independently in two different settings. Lagrange (Lagrange 1770–1773) used the recurrence relation

$$(3) \qquad \begin{aligned} (2n+1)xP_n(x) &= (n+1)P_{n+1}(x) + nP_{n-1}(x), \\ &\qquad n = 1, 2, \ldots, \ P_0(x) = 1, \ P_1(x) = x. \end{aligned}$$

The second setting was work by Legendre and Laplace on celestial mechanics. Here the generating function

$$(4) \qquad (1 - 2xr + r^2)^{-1/2} = \sum_{n=0}^{\infty} P_n(x) r^n$$

and variants of it are one of the main reasons Legendre polynomials appear in this work. Many papers could be cited. The early ones are mentioned in (Heine 1878). (See also Chapter 5.)

Rodrigues came a generation later. Rodrigues's 1815 thesis was published in 1816 in a journal which had only three volumes. That partly explains why (Rodrigues 1816) was overlooked for so long. Another identity related to (2) which was contained in this paper of Rodrigues was

$$(5) \qquad \frac{1}{(n-k)!} \frac{d^{n-k}}{dx^{n-k}} (x^2-1)^n = \frac{(x^2-1)^k}{(n+k)!} \frac{d^{n+k}}{dx^{n+k}} (x^2-1)^n, \qquad k = 1, 2, \ldots.$$

Formula (5) was rediscovered by Jacobi (Jacobi 1827). Another result similar to (2) was published in (Jacobi 1836):

$$(6) \qquad \frac{d^n}{dx^n}(1-x^2)^{n+\frac{1}{2}} = (-1)^n \, 3 \cdot 5 \cdots (2n+1) \frac{\sin(n+1)\theta}{n+1}$$

where $x = \cos\theta$. He then concluded that

$$(7) \qquad \frac{d^n}{dx^n}(1-x^2)^{n-\frac{1}{2}} dx = (-1)^{n-1} \, 3 \cdot 5 \cdots (2n-1) \cos n\theta d\theta$$

and also

$$(8) \qquad \int_0^{\pi} f^{(n)}(\cos\theta)(\sin\theta)^{2n} d\theta = 3 \cdot 5 \cdots (2n-1) \int_0^{\pi} f(\cos\theta) \cos n\theta d\theta.$$

In a posthumous paper, Jacobi (Jacobi 1859) proved a result equivalent to

$$(9) \qquad (1-x)^{\alpha}(1+x)^{\beta} P_n^{(\alpha,\beta)}(x) = \frac{(-1)^n}{2^n n!} \frac{d^n}{dx^n}[(1-x)^{n+\alpha}(1+x)^{n+\beta}].$$

The polynomials $P_n^{(\alpha,\beta)}(x)$ are called Jacobi polynomials. They can be represented as

$$P_n^{(\alpha,\beta)}(x) = \frac{(\alpha+1)}{n!}\,{}_2F_1\left(\begin{matrix}-n, n+\alpha+\beta+1\\ \alpha+1\end{matrix}; \frac{1-x}{2}\right) \tag{10}$$

where

$$\begin{aligned}(a)_n &= a(a+1)\cdots(a+n-1), && n = 1,3,\dots,\\ &= 1, && n = 0,\end{aligned} \tag{11}$$

and

$${}_pF_q\left(\begin{matrix}a_1,\dots,a_p\\ b_1,\dots,b_q\end{matrix}; x\right) = \sum_{k=0}^{\infty} \frac{(a_1)_k \cdots (a_p)_k}{(b_1)_k \cdots (b_q)_k} \frac{x^k}{k!}. \tag{12}$$

When $\alpha, \beta > -1$, Jacobi polynomials are orthogonal on $(-1,1)$:

$$\begin{aligned}\int_{-1}^{1} P_n^{(\alpha,\beta)}(x) P_m^{(\alpha,\beta)}(x)(1-x)^\alpha(1+x)^\beta dx &= 0, \qquad m \neq n,\\ &= \frac{2^{\alpha+\beta+1}\Gamma(n+\alpha+1)\Gamma(n+\beta+1)}{(2n+\alpha+\beta+1)n!\Gamma(n+\alpha+\beta+1)}.\end{aligned} \tag{13}$$

The orthogonality relation (13) follows easily from (9) by integration by parts.

When Hermite wrote (Hermite 1865*a*) and (Hermite 1865*b*), he knew about Jacobi (Jacobi 1827, 1836) but did not seem to know about (Jacobi 1859). Among other results Hermite found two variable analogues of (2) and (6) for

$$\frac{d^n}{dx^k dy^{n-k}}(1-x^2-y^2)^{n+j} \tag{14}$$

when $j = 0$ and $j = \frac{1}{2}$. See (Hermite 1865*a*) for $j = 0$ and (Hermite 1865*b*) for $j = \frac{1}{2}$. See (Appell and Kampé de Fériet 1926) for more information on these polynomials and extensions to k variables.

Extensions to other orthogonal polynomials

There are two other sets of orthogonal polynomials with Rodrigues type formulas involving an nth derivative. These are Hermite polynomials, $H_n(x)$, and Laguerre polynomials, $L_n^\alpha(x)$. These formulae are

$$e^{-x^2} H_n(x) = (-1)^n \frac{d^n}{dx^n} e^{-x^2}, \tag{15}$$

$$x^\alpha e^{-x} L_n^\alpha(x) = \frac{1}{n!}\frac{d^n}{dx^n}\left[x^{n+\alpha}e^{-x}\right]. \tag{16}$$

Laplace (Laplace 1810) found Hermite polynomials, and Abel ((Abel 1881) but done much earlier) and Murphy (Murphy 1833) found Laguerre polynomials (when $\alpha = 0$) long before Hermite and Laguerre rediscovered them. See (Shohat, Hille, and Walsh 1940) for many other references and Szegö (Szegö 1938–1975) for the above formulas for Jacobi, Laguerre, and Hermite polynomials and many other important formulae.

Tricomi (Tricomi 1948) showed that if

$$w(x)p_n(x) = \frac{d^n}{dx^n}\left[w(x)\{R(x)\}^n\right], \qquad n = 0, 1, \dots, \tag{17}$$

where $R(x)$ is a fixed polynomial and $p_n(x)$ is a polynomial of degree n, then $\{p_n(x)\}$ can be rescaled to be Jacobi, Laguerre, or Hermite polynomials or limiting cases of Jacobi polynomials such as Bessel polynomials or x^n. The only ones which are orthogonal with respect to a positive measure are Jacobi, Laguerre, and Hermite polynomials when $\alpha, \beta > -1$ for Jacobi and $\alpha > -1$ for Laguerre, when the orthogonality exists for all $n = 0, 1, \ldots$. For Jacobi and Laguerre polynomials there can be orthogonality with respect to a positive measure for a finite number of polynomials when these conditions are not satisfied.

However, there are other extensions of Rodrigues's formula (2) which lead to orthogonal polynomials. The first person to realize this was Tchebychef (Tchebychef 1858). His name is now transliterated as Chebyshev, and this form will be used from now on.

Define

$$\Delta_h f(x) = f(x+h) - f(x). \tag{18}$$

Later we will take $h = 1$ and just use $\Delta = \Delta_1$.

Chebyshev's (Tchebyshev 1858) analogue of (2) is

$$\Delta_h^n (x-h)(x-2h)\cdots(x-nh)(x-(N+1)h)(x-(N+2)h)\cdots(x-(N+n)h). \tag{19}$$

These polynomials generalize Legendre polynomials, as Chebyshev (Tchebyshev 1858) observed. He also stated

$$\begin{aligned}\sum_{i=1}^{N} u_i^2 = \frac{(\Sigma u_i)^2}{N} &+ \frac{3(\Sigma i(N-i)\Delta_h u_i)^2}{1^2 N(N^2-1^2)h^2} \\ &+ \frac{5(\Sigma i(i+1)(N-i)(N-i-1)\Delta_h^2 u_i)^2}{1^2\cdot 2^2 N(N^2-1^2)(N^2-2^2)h^4} \\ &+ \frac{7(\Sigma i(i+1)(i+2)(N-i)(N-i-1)(N-i-2)\Delta_h^3 u_i)^2}{1^2\cdot 2^2\cdot 3^2 N(N^2-1)(N^2-2^2)(N^2-3^2)h^6} + \cdots.\end{aligned} \tag{20}$$

Chebyshev (Tchebyshev 1858) wrote that when $h = N^{-1}$ goes to zero, this becomes

$$\begin{aligned}\int_0^1 [u(x)]^2 dx = &\left(\int_0^1 u(x)dx\right)^2 + \frac{3}{1^2}\left(\int_0^1 x(1-x)\frac{du}{dx}dx\right)^2 \\ &+ \frac{5}{1^2\cdot 2^2}\left(\int_0^1 x^2(1-x)^2\frac{d^2u}{dx^2}dx\right)^2 \\ &+ \frac{7}{1^2\cdot 2^2\cdot 3^2}\left(\int_0^1 x^3(1-x)^3\frac{d^3u}{dx^3}dx\right)^3 + \cdots.\end{aligned} \tag{21}$$

Formula (21) can be proved directly when (2) is modified, replacing $1-x^2$ by $x(1-x)$, and orthogonality is used. A similar proof gives (20) after the orthogonality of the polynomials in (19) has been worked out. Chebyshev clearly knew this but did not state it in this paper. Later he gave this in (Tchebychef 1864).

Chebyshev (Tchebyshev 1875) found similar polynomials which generalize Jacobi polynomials. He gave a finite differences analogue of (9), the orthogonality relation and an explicit series for these polynomials. However, this paper was not appreciated and was overlooked or ignored, so his results had to be rediscovered.

These polynomials are now called Hahn polynomials and are denoted by

$$Q_n(x) = Q_n(x;\alpha,\beta,N) = {}_3F_2\left(\begin{matrix}-n, n+\alpha+\beta+1, -x\\ \alpha+1, -N\end{matrix};1\right), \tag{22}$$

$x, n = 0, 1, \ldots, N$. The orthogonality relation is

$$\sum_{x=0}^{N} Q_n(x)Q_m(x)\frac{(\alpha+1)_x}{x!}\frac{(\beta+1)_{N-x}}{(N-x)!} = 0, \qquad m \neq n \leq N. \tag{23}$$

The weight function can also be written as $\binom{x+\alpha}{x}\binom{N-x+\beta}{N-x}$.

The analogue of Rodrigues's formula is

(24)

$$\Delta^n \binom{x+\alpha}{x-n}\binom{N-x+\beta+n}{N-x} = k_n \binom{x+\alpha}{x}\binom{N-x+\beta}{N-x} Q_n(x;\alpha,\beta,N)$$

where $k_n = k_n(\alpha,\beta,N)$ is independent of x.

The value of (23) when $m = n$ and the value of k_n are known; see (Koekoek and Swarttouw 1998). A formula equivalent to (24) was derived by Weber and Erdélyi (Weber and Erdélyi 1952) when they proved a finite difference analogue of Tricomi's result (17). There are other sets of orthogonal polynomials which satisfy a Rodrigues type formula like (24). The weight functions are the binomial, negative binomial, and Poisson distributions. There are others, a continuous version of a binomial distribution and of the hypergeometric distribution. See (Koekoek and Swarttouw 1998) for these polynomials.

However, there is more, even for polynomials that can be given as hypergeometric series, which is the formal name for the functions defined in (12). The most general classical type orthogonal polynomials which can be represented as a hypergeometric series occur at the ${}_4F_3$ level. There are two cases, when the orthogonality relation is discrete on a finite set of points, as was the case for (22) and (23), and when it is absolutely continuous with a possible finite number of discrete mass points. See (Wilson 1980) for this case. The formulae in the absolutely continuous case without discrete masses are the most attractive, so only this case will be stated here.

The Wilson polynomials are polynomials of even degree in x but of general degree in x^2:

(25)

$$\frac{W_n(x^2;a,b,c,d)}{(a+b)_n(a+c)_n(a+d)_n} = {}_4F_3\left(\begin{matrix}-n, n+a+b+c+d-1, a+ix, a-ix\\ a+b, a+c, a+d\end{matrix};1\right).$$

When $a, b, c, d > 0$, the orthogonality relation is

(26)

$$\int_0^\infty W_n(x^2)W_m(x^2)\left|\frac{\Gamma(a+ix)\Gamma(b+ix)\Gamma(c+ix)\Gamma(d+ix)}{\Gamma(2ix)}\right|^2 dx = 0,\ m \neq n.$$

To motivate the Rodrigues type result, recall that when $p_n(x)$ is a polynomial of degree n in x, then $p_n(x^2)$ is a polynomial of degree n in x^2, and

$$\frac{1}{x}\frac{d}{dx}p_n(x^2) = 2p_n'(x^2) = q_{n-1}(x^2) \tag{27}$$

is a polynomial of degree $n-1$ in x^2. Wilson's discrete analogue of (27) is the divided difference operator

$$\nabla p_n(x^2) = \frac{p_n\left(\left(x+\frac{i}{2}\right)^2\right) - p_n\left(\left(x-\frac{i}{2}\right)^2\right)}{\left(x+\frac{i}{2}\right)^2 - \left(x-\frac{i}{2}\right)^2}. \tag{28}$$

The Rodrigues type formula is

$$\nabla^n w\left(x; a+\frac{n}{2}, b+\frac{n}{2}, c+\frac{n}{2}, d+\frac{n}{2}\right) = k_n W_n(x^2; a,b,c,d) w(x,a,b,c,d) \tag{29}$$

where

$$w(x;a,b,c,d) = \left|\frac{\Gamma(a+ix)\Gamma(b+ix)\Gamma(c+ix)\Gamma(d+ix)}{\Gamma(2ix)}\right|^2$$

and k_n is independent of x.

Before this example, each of the Rodrigues type formulas used the weight function, possibly with part of it shifted in x, times an appropriate polynomial. Now we see that this was misleading. The polynomial comes from shifting the parameters in the weight function.

The Rodrigues paper of 1839

There will be other operators and polynomials, but we turn to the second result of Rodrigues given in his paper of 1839. Cramer's work on determinants led Stern (Stern 1838) to ask the following question:

Take the integers $\{1, 2, \ldots, n\}$ and consider the $n!$ permutations of them. For each permutation, consider the number of inversions. An inversion occurs when a larger number occurs between a smaller number. For example $\{4, 2, 1, 3\}$ has four inversions since 4 appears before $2, 1$, and 3 and since 2 appears before 1. Stern (Stern 1838) asked how many inversions there are among these $n!$ permutations. The answer, $n! \cdot n(n-1)/4$, was shown by Terquem (Terquem 1838). Rodrigues (Rodrigues 1839) solved this problem by solving a more general and more important problem. He found a generating function for these inversions. If $I(n,k)$ denotes the number of permutations of $\{1, 2, \ldots, n\}$ with k inversions, then

$$\sum_{k\geq 0} I(n,k)q^k = 1(1+q)(1+q+q^2)\cdots(1+q+q^2+\cdots+q^{n-1}). \tag{30}$$

Since the case $q=1$ gives $n!$, the function on the right is called a q-factorial and can be denoted by

$$n!_q = 1(1+q)\cdots(1+q+\cdots+q^{n-1}). \tag{31}$$

Once a generalized factorial has been introduced, it is natural to ask if there are generalized binomial coefficients and a generalization of the binomial theorem which is a generating function of these generalized binomial coefficients. Gauss (Gauss 1811) and Rothe (Rothe 1811) both introduced

$$\begin{bmatrix} n \\ k \end{bmatrix}_q = \frac{n!_q}{k!_q(n-k)!_q}, \tag{32}$$

for different reasons. Rothe (Rothe 1811) stated a q-version of the binomial theorem. Correcting some missing letters which were dropped in the printing, he stated that

$$(x+y)(x+qy)\cdots(x+q^{n-1}y)=\sum_{k=0}^{n}\begin{bmatrix}n\\k\end{bmatrix}_q q^{k(k-1)/2}x^{n-k}y^k. \tag{33}$$

This formula is only stated in the foreword as an example of what he had to omit since the publisher felt the original manuscript was too long. Gauss later rediscovered (33), but he did not publish it, so (33) was not stated and proved in a journal until it was rediscovered by Cauchy (Cauchy 1843) and Heine (Heine 1847).

There is a connection with Rodrigues's generating function (30), but the version of (33) using inversions would only be found much later.

MacMahon (MacMahon 1916) seems to have been the first to look seriously at inversions of multisets. Instead of considering the set $\{1,2,\ldots,n\}$, he considered $\{1^{j_1},2^{j_2},\ldots,n^{j_n}\}$ where 1 occurs j_1 times, etc. When $n=2$, the generating function is the q-binomial coefficient (32) with $j_1=k$, $j_2=n-k$. See Andrews's comments in Chapter 4 of (MacMahon 1978) and the papers of MacMahon reprinted in that book for other partition statistics.

The first version of a q-binomial theorem with a twisted multiplication seems to be Porter (Porter 1950). Olga Holtz pointed this out recently. Shortly after this, Schützenberger (Schützenberger 1953) gave an equivalent result. This is

$$(x+y)^n=\sum_{k=0}^{n}\begin{bmatrix}n\\k\end{bmatrix}_q x^{n-k}y^k \tag{34}$$

when $yx=qxy$, $qx=xq$, $qy=yq$.

This makes the connection with inversions explicit. Formula (34) can be iterated to give a multinomial versions of (34) which contains (30) as the special case when the coefficient of $x_1x_2\cdots x_n$ is determined in the expansion of $(x_1+x_2+\cdots+x_n)^n$ when $x_ix_j=qx_jx_i$ for $1\le i<j\le n$, $qx_i=x_iq$, $i=1,2,\ldots,n$.

A lot more could be written about inversions. A recent paper is (Ding and Wu 2003). Probably the most important use of the noncommutative q-binomial theorem is in some specific quantum groups. The connection between the two results of Rodrigues comes from some orthogonal polynomials which arise in this setting. We now return to orthogonal polynomials to describe some q-versions of Rodrigues's formula (2).

Probably the first person to find a set of orthogonal polynomials which comes from q-series, or basic hypergeometric series, was Markoff (Markoff 1884). He found a q-extension of the first discrete polynomials (19) found by Chebyshev.

Basic hypergeometric series are an extension of hypergeometric series. Hypergeometric series are series whose term ratio is a rational function of n. Basic hypergeometric series have a term ratio which is a rational function of q^n for a fixed q:

$${}_r\varphi_s\left(\begin{matrix}a_1,\ldots,a_r\\b_1,\ldots,b_s\end{matrix};q,x\right)=\sum_{k=0}^{\infty}c_kx^k \tag{35}$$

where $c_0=1$,

$$\frac{c_{k+1}}{c_k}=\frac{(1-a_1q^k)\cdots(1-a_rq^k)(-q^k)^{(s+1-r)}}{(1-b_1q^k)\cdots(1-b_sq^k)(1-q^{k+1})}. \tag{36}$$

This can be uncoupled using

$$(37)\qquad \begin{aligned}(a;q)_k = &\quad (1-a)(1-aq)\cdots(1-aq^{k-1}), && k=1,2,\dots\\ =1 &&& k=0,\end{aligned}$$

to give

$$(38)\qquad {}_r\varphi_s\left(\begin{matrix}a_1,\dots,a_r\\ b_1,\dots,b_x\end{matrix};q,x\right) = \sum_{k=0}^{\infty}\frac{(a_1;q)_k\cdots(a_r;q)_k(-1)^{(s+1-r)k}}{(b_1;q)_k\cdots(b_s;q)_k(q;q)_k}q^{\binom{k}{2}(s+1-r)}x^k.$$

Stieltjes (Stieltjes 1894) and Rogers (Rogers 1893*a*, 1893*b*, 1894, 1895) then found others. Stieltjes' polynomials are one of many q-extensions of Hermite polynomials. Rogers first found a different q-extension of Hermite polynomials and in (Rogers 1895) found a set of polynomials which extends the symmetric Jacobi polynomials when $\alpha = \beta$. Rogers was unaware that his polynomials were orthogonal, but he knew they were important.

Later extensions

In the first half of the twentieth century other q-extensions of the classical orthogonal polynomials were found, but they seem isolated and not part of a larger picture. The person who first tied all of this together was Hahn (Hahn 1949).

Hahn considered the operator

$$(39)\qquad D_q f(x) = \frac{f(qx)-f(x)}{qx-x}.$$

He raised and answered a number of questions about when the derivative operator is replaced by D_q. One is a q-extension of the following result for Jacobi, Laguerre, and Hermite polynomials. It is easy to see that the derivative of a polynomial in one of these three families has the property of also being in the same family with (α,β) changed to $(\alpha+1,\beta+1)$ for Jacobi and α to $\alpha+1$ for Laguerre. These are the only families of orthogonal polynomials $\{p_n(x)\}$ for which $\{p'_{n+1}(x)\}$ is another set of orthogonal polynomials if polynomials which can be mapped to each other by $x \to ax+b$ are identified. Hahn found the corresponding sets of orthogonal polynomials $\{p_n(x)\}$ for which $\{D_q p_{n+1}(x)\}$ is a set of orthogonal polynomials. He showed that the same set of polynomials arises when a few other conditions are considered. One of these conditions was a Rodrigues type formula with D_q used instead of the derivative. He carried out the details, including finding the orthogonality relation for one of these families. These are a class of q-Jacobi polynomials:

$$(40)\qquad p_n(x;\alpha,\beta:q) = {}_2\varphi_1\left(\begin{matrix}q^{-n},q^{n+\alpha+\beta+1}\\ q^{\alpha+1}\end{matrix};q,qx\right).$$

Hahn's orthogonality relation is a q-extension of (13) after (13) is put on $(0,1)$. The measure uses what I call Fermat measure since Fermat used this type of measure to integrate x^n:

$$(41)\qquad \int_0^a f(t)d_qt = a(1-q)\sum_{k=1}^{\infty} f(aq^k)q^k.$$

A q-analogue of Euler's beta function integral

$$(42)\qquad \int_0^1 t^{a-1}(1-t)^{b-1}dt = \frac{\Gamma(a)\Gamma(b)}{\Gamma(a+b)}$$

is

$$\int_0^1 t^{a-1}\frac{(tq;q)_\infty}{(tq^b;q)_\infty}d_qt = \frac{\Gamma_q(a)\Gamma_q(b)}{\Gamma_q(a+b)} \tag{43}$$

where for simplicity we take $0 < q < 1$ and define

$$(a;q)_\infty = \prod_{k=0}^{\infty}(1-aq^k), \tag{44}$$

$$\Gamma_q(a) = \frac{(q;q)_\infty(1-q)^{1-a}}{(q^a;q)_\infty}. \tag{45}$$

It is easy to see that

$$\Gamma_q(n+1) = n!_q, \tag{46}$$

so $\Gamma_q(a)$ is a natural extension of the q-factorial to noninteger values.

Hahn's orthogonality is

$$\int_0^1 p_n(x;\alpha,\beta:q)p_m(x;\alpha,\beta:q)w(x)d_qx = 0, \qquad m \neq n, \tag{47}$$

when $\alpha > -1$, $\beta > -1$, and

$$w(x) = w(x;\alpha,\beta:q) = x^\alpha\frac{(qx;q)_\infty}{(q^{\beta+1}x;q)_\infty}. \tag{48}$$

The Rodrigues type formula for these polynomials is

$$w(x;\alpha,\beta:q)p_n(x;\alpha,\beta:q) = \frac{q^{n\alpha+\binom{n}{2}}(1-q)^n}{(q^\alpha;q)_n}, \quad D^n_{q^{-1}}[w(x;\alpha+n,\beta+n:q)]. \tag{49}$$

The most general set of polynomials which arises in Hahn's theorem is

$$Q_n(x;a,b,c:q) = {}_3\varphi_2\left(\begin{matrix} q^{-n}, q^{n+1}ab, x \\ aq, cq \end{matrix}; q, q\right). \tag{50}$$

When $cq = q^{-N}$, there is an orthogonality relation extending (23). For different conditions there is a different orthogonality extending (13) which is the sum of two integrals using Fermat measure, one with the mass points approaching 0 through positive values and the other through negative values. (See (Andrews and Askey 1985).)

As in the hypergeometric case, this is not the full story. The polynomials of (Rogers 1893*a*, 1895) are not contained in Hahn's theorem. These are such nice polynomials that one should look for a setting where they arise. One was found by extending Wilson's operator (28).

Rogers's polynomials extend the symmetric Jacobi polynomials with a different normalization. These polynomials are called ultraspherical polynomials. A generating function is

$$\sum_{n=0}^{\infty} C_n^\lambda(x)r^n = (1-2xr+r^2)^{-\lambda}. \tag{51}$$

Differentiation with respect to x gives

$$\frac{d}{dx}C_n^\lambda(x) = 2\lambda C_{n-1}^{\lambda+1}(x). \tag{52}$$

Their three term recurrence relation is

$$2(n+\lambda)xC_n^\lambda(x) = (n+1)C_{n+1}^\lambda(x) + (n+2\lambda-1)C_{n-1}^\lambda(x). \tag{53}$$

The orthogonality relation is

$$\int_{-1}^{1} C_n^\lambda(x)C_m^\lambda(x)(1-x^2)^{\lambda-\frac{1}{2}}dx = 0, \qquad m \neq n,\ \lambda > -\frac{1}{2}. \tag{54}$$

Rogers's polynomials satisfy

$$\begin{aligned} 2(1-\beta q^n)xC_n(x;\beta \mid q) &= (1-q^{n+1})C_{n+1}(x;\beta \mid q) \\ &\quad + (1-\beta^2 q^{n-1})C_{n-1}(x; l\beta \mid q), \end{aligned} \tag{55}$$

$$C_0(x;\beta \mid q) = 1,\ C_1(x;\beta \mid q) = 2(1-\beta)x/(1-q).$$

When $\beta = q^\lambda$ and (55) is divided by $1-q$, the result of letting $q \to 1$ is (53).

The corresponding generating function is

$$\prod_{k=0}^{\infty} \frac{(1-2xr\beta q^k + r^2\beta^2 q^{2k})}{(1-2xrq^k + r^2q^{2k})} = \sum_{n=0}^{\infty} C_n(x;\beta \mid q)r^n. \tag{56}$$

This can be changed to look more like (51) if $x = \cos\theta$ is used in both. The function (51) is

$$(1-re^{i\theta})^{-\lambda}(1-re^{-\theta})^{-\lambda}. \tag{57}$$

The function in (56) becomes

$$\frac{(\beta re^{i\theta};q)_\infty(\beta re^{-i\theta};q)_\infty}{(re^{i\theta};q)_\infty(re^{-i\theta};q)_\infty}. \tag{58}$$

The connection comes from an infinite form of the binomial theorem and the q-binomial theorem. These are

$$(1-r)^{-\lambda} = \sum_{n=0}^{\infty} \frac{(\lambda)_n}{n!}r^n, \tag{59}$$

$$\frac{(\beta r;q)_\infty}{(r;q)_\infty} = \sum_{n=0}^{\infty} \frac{(\beta;q)_n}{(q;q)_n}r^n. \tag{60}$$

Once one has this generating function for Rogers's polynomials, which he had, it is easy to find an operator which acts on (58) in the same way the derivative acts on (51).

If $f(x) = f(\cos\theta) = F(e^{i\theta})$, then

$$\delta_q f(x) = F(q^{1/2}e^{i\theta}) - F(q^{-1/2}e^{i\theta}) \tag{61}$$

and

$$\lim_{q\to 1} \frac{\delta_q f(x)}{\delta_q x} = \frac{df(x)}{dx}. \tag{62}$$

On using the operator

$$D_q f(x) = \frac{\delta_q f(x)}{\delta_q x}, \tag{63}$$

it is possible to give a Rodrigues type formula for these polynomials of Rogers. However, one can go beyond these polynomials of Rogers to a four-parameter set which generalizes the polynomials (25) of Wilson. These polynomials are

$$\frac{a^n W_n(\cos\theta; a, b, c, d \mid q)}{(ab;q)_n(ac;q)_n(ad;q)_n} = {}_4\varphi_3\left(\begin{matrix} q^{-n}, q^{n-1}abcd, ae^{i\theta}, ae^{-i\theta} \\ ab, ac, ad \end{matrix}; q, q\right). \tag{64}$$

When $-1 < a, b, c, d < 1$ and $m \neq n$, then

$$\int_0^\pi W_n(\cos\theta)W_m(\cos\theta)\left|\frac{(e^{2i\theta};q)_\infty}{(ae^{i\theta};q)_\infty(be^{i\theta};q)_\infty(ce^{i\theta};q)_\infty(de^{i\theta};q)_\infty}\right|^2 d\theta = 0. \tag{65}$$

Just as for the Wilson polynomials, the Rodrigues type relation involves a shift in a, b, c, d in the weight function. Each is replaced by itself times $q^{n/2}$. The resulting formula is given in (Askey and Wilson 1985).

The binomial theorem and q-binomial theorem provide one connection between the two results of Rodrigues, after they have been generalized to Jacobi polynomials in the first case and q-binomial theorems in the second. However, there are further connections which could be said to have been hinted at in the second edition of (Heine 1878). This is a book about spherical harmonics, and Rodrigues's result (2) and the extensions through Jacobi polynomials fit in nicely. However, in his second edition, Heine included a section on basic hypergeometric series. Probably the main reason was that he had done some very nice work on ${}_2\varphi_1$'s thirty years earlier and it had largely been ignored, so he wanted to call attention to it in a book which would be read by many more people than would read a thirty-year-old paper. However, he knew that his q-series results were natural extensions of some of the results in his book. He succeeded to some extent, since Rogers read this material and cited it in the first sentence in (Rogers 1893*a*), which was the first of his series of four papers dealing with some polynomials and what could be done with them.

A structural condition on a divided difference operator which follows from it, taking a polynomial of degree n in $\mu(x)$ to one of degree $n-1$ in $\mu(x)$, was found by Magnus (Magnus 1988). The operator

$$\frac{\delta_q f(x)}{\delta_q x}$$

is basically a symmetrized example of the general Magnus operator.

There is a much deeper connection than Heine could have been aware of. From work of Legendre and Laplace it was clear that rotations of the sphere were important. After group theory was developed, it was clear that the rotation group acting on the sphere could be extended to some other settings. One setting is rank one compact symmetric spaces. See (Helgason 1984) for the orthogonal polynomials which arise, certain Jacobi polynomials. A second setting is representation theory of $SU(2)$. See (Vilenkin and Klimyk 1991) for the special functions which arise there.

In the past twenty or so years q-versions of $SU(2)$ and of spheres have been discovered. The corresponding special functions which arise contain many of the q-orthogonal polynomials mentioned above. These include the discrete version of the ${}_4\varphi_3$ polynomials. The orthogonality in this case was given by Askey and Wilson (Askey and Wilson 1979). They did not state a Rodrigues type formula for these polynomials, but one exists and can be found by translating the one they gave in (Askey and Wilson 1985). For details of $SU_q(2)$ see (Koelink 1996).

Both Rodrigues and Heine would likely have been surprised by these developments. I was.

Bibliography

Abel, N. H. (1881). Sur une espèce particulière de fonctions entières nées du développement de la fonction $(1-v)^{-1}e^{-xv/(1-v)}$ suivant les puissances de v. *Oeuvres complètes de Niels Henrik Abel*, Christiania, **2**, 284.

Andrews, G. E. and R. Askey (1985). Classical orthogonal polynomials, in *Polynômes Orthogonaux et Applications* (C. Brezinski et al., eds.), pp. 36–62. Springer-Verlag, Berlin.

Appell, P. and J. Kampé du Fériet (1926). *Fonctions hypergéométriques et hypersphériques. Polynomes d'Hermite.* Gauthier-Villars, Paris.

Askey, R. and J. Wilson (1979). A set of orthogonal polynomials that generalize the Racah coefficients or $6-j$ symbols. *SIAM Journal on Mathematical Analysis*, **10**, 1008–1016.

Askey, R. and J. Wilson (1985). Some basic hypergeometric orthogonal polynomials that generalize Jacobi polynomials. *Memoirs American Mathematical Society*, **54**, no. 319.

Carlitz, L. (1970). Sequences and inversions. *Duke Mathematical Journal*, **37**, 193–198.

Cauchy, A. (1843). Mémoire sur les fonctions dont plusieurs valeurs sont liées entre elles par une équation linéaire, et sur diverses transformations de produits composés d'un nombre indéfini de facteurs. *Comptes Rendus de l'Académie des Sciences*, Paris, **17**, 523. Also *Oeuvres Complétes d'Augustin Cauchy*, (1893), **8**, 1st series, pp. 42–50. Gauthier-Villars, Paris.

Cayley, A. (1858). On Rodrigues' method for the attraction of ellipsoids. *Quarterly Journal of Pure and Applied Mathematics*, **2**, 333–337.

Chebyshev, P. L. (see Tchebychef, P. L.)

Ding, K. and S. Wu (2003). Inversions in classical Weyl groups. Preprint.

Gauss, C. F. (1811). Summatio quarundam serierum singularium. Commentationes societatis regine scientiarum Gottingensis recentiores, 1. *Werke*, **2**, Göttingen (1863), 11–45.

Hahn, W. (1949). Über Orthogonalpolynome, die q-Differenzengleichungen genügen. *Mathematische Nachrichten*, **2**, 4–34.

Heine, E. (1847). Untersuchungen über die Reihe $1+\frac{(1-q^\alpha)(1-q^\beta)}{(1-q)(1-q^\gamma)}x+\frac{(1-q^\alpha)(1-q^{\alpha+1})(1-q^\beta)(1-q^{\beta+1})}{(1-q)(1-q^2)(1-q^\gamma)(1-q^{\gamma+1})}x^2+\ldots$, *Journal für die Reine und Angewandte Mathematik*, **34**, 285–328.

Heine, E. (1861). *Handbuch der Kugelfunctionen.* Reimer, Berlin.

Heine, E. (1878). *Theorie der Kugelfunctionen und der verwandten Functionen.* Reimer, Berlin.

Helgason, S. (1984). *Groups and Geometric Analysis.* Academic Press, Orlando.

Hermite, C. (1865*a*). Extrait d'une letter de M. Hermite à M. Borchardt. *Journal für Reine and Angewandte Mathematik*, 294–297. Also *Oeuvres de Charles Hermite*, (1918), **2**, pp. 309–312. Paris.

Hermite, C. (1865*b*). Sur quelques développements en série de fonctions de plusieurs variables. Series of notes in *Comptes Rendus*, Paris, **60**. Also *Oeuvres de Charles Hermite*, (1918), **2**, pp. 319–346, Paris.

Ivory, J. (1812). On the attractions of an extensive class of spheroids. *Philosophical Transactions of the Royal Society*, London, **102**, 46–82.

Ivory, J. (1824). On the figure requisite to maintain the equilibrium of a homogeneous fluid mass that revolves upon an axis. *Philosophical Transactions of the Royal Society*, London, **114**, 85–150.

Ivory, J. and C. G. J. Jacobi (1837). Sur le developpement de $(1-2xz+z^2)^{-1/2}$. *Journal de Mathématiques Pures et Appliquées* (1) **2**, 105–106.

Jacobi, C. G. J. (1827). Über eine besondere Gattung algebraischer Functionen, die aus der Entwickelung der Function $(1-2xz+z^2)^{1/2}$ entstehen. *Journal für die Reine und Angewandte Mathematik*, **2**, 223–226.

Jacobi, C. G. J. (1836). Formula transformationis integralium definitorum. *Journal für die Reine und Angewandte Mathematik*, **15**, 1–26.

Jacobi, C. G. J. (1859). Untersuchungen über die Differentialgleichung der hypergeometrischen Reihe. *Journal für die Reine und Angewandte Mathematik*, **56**, 149–165.

Koekoek, R. and R. Swarttouw (1998). The Askey scheme of hypergeometric orthogonal polynomials and its q-analogue. Report 98–17, Department of Technical Mathematics and Informatics, Delft University of Technology, Delft. `http://aw.twi.tudelft.nl/~koekoek/askey`

Koelink, H. T. (1996). Askey-Wilson polynomials and the quantum $SU(2)$ group. Survey and applications. *Acta Applicandae Mathematicae*, **44**, 295–352.

Lagrange, J. (1770–1773). Mémoire sur l'utilité de la methodé de prendre le milieu. *Misc. Taurinensia*, **5**. Also *Oeuvres de Lagrange* (J.-A. Serret, ed.), (1868), pp. 173–234. Gauthier-Villars, Paris.

Laplace, P. S. (1810). Mémoire sur les intégrales définies, et leur application aux probabilités. *Mémoires de l'Académie des Sciences*, Ire Série, **11**, (Ire Partie). Also *Oeuvres Complètes de Laplace*, (1898), **12**. Gauthier-Villars, Paris.

MacMahon, P. A. (1916). Two applications of general theorems in combinatory analysis: (1) to the theory of inversions of permutations; (2) to the ascertainment of the numbers of terms in the development of a determinant which has amongst its elements an arbitrary number of zeros. *Proceedings of the London Mathematical Society* (2), **15**, 314–321.

MacMahon, P. A. (1978). Collected Papers, Volume I, Combinatorics (E. Andrews, ed.), MIT Press, Cambridge, MA and London.

Magnus, A. (1988). Associated Askey-Wilson polynomials as Laguerre Hahn orthogonal polynomials. In *Orthogonal Polynomials and Their Applications*, pp. 261–278. Springer-Verlag, Berlin.

Markoff, A. (1884). *On some applications of algebraic continued fractions* (in Russian). Thesis, 131 pp., St. Petersburg.

Murphy, R. (1833). On the inverse method of definite integrals, with physical applications. *Transactions of the Cambridge Philosophical Society*, **4**, 353–408; (1835), **5**, 113–148, 315–393.

Potter, H. S. A. (1950). On the latent roots of quasi-commutative matrices. *The American Mathematical Monthly*, **57**, 321–322.

Rodrigues, O. (1816). Mémoire sur l'attraction des sphéroides. *Correspondence sur l'École Polytechnique*, **3**, 361–385.

Rodrigues, O. (1839). Note sur les inversions, ou dérangements produits dans les permutations. *Journal de Mathématiques Pures et Appliquées*, **4**, 236–240.

Rogers, L. J. (1893*a*). On a three-fold symmetry in the elements of Heine's series. *Proceedings of the London Mathematical Society*, **24**, 171-179.

Rogers, L. J. (1893*b*). On the expansion of some infinite products. *Proceedings of the London Mathematical Society*, **24**, 337–352.

Rogers, L. J. (1894). Second memoir on the expansion of certain infinite products. *Proceedings of the London Mathematical Society*, **25**, 318–343.

Rogers, L. J. (1895). Third memoir on the expansion of certain infinite products. *Proceedings of the London Mathematical Society*, **26**, 15–32.

Rothe, H. A. (1811). *Lehrbuch der Arithmetic*. Leipzig.

Schützenberger, M.-P. (1953). Une interprétation de certaines solutions de l'equation fonctionnelle: $F(x+y) = F(x)F(y)$. *Comptes Rendus de l'Académie des Sciences*, Paris, **236**, 352–353.

Shohat, J. A., E. Hille, and J. L. Walsh (1940). *A Bibliography on Orthogonal Polynomials*. Bulletin of the National Research Council, **103**, National Academy of Sciences, Washington, DC.

Stern, M. (1838). Aufgaben. *Journal für die Reine und Angewandte Mathematik*, **18**, 100.

Stieltjes, T. J. (1894). Recherches sur les fractions continues. *Annales de la Faculté des Sciences de Toulouse*, **8**, 122 pp. (1895), **9**, 47 pp. Also *Oeuvres Complètes*, **2**, 402–566.

Szegö, G. (1938–1975). *Orthogonal Polynomials*. American Mathematical Society.

Tchebychef, P. L. (1858). Sur une nouvelle série. *Bulletin Physico-Mathématique de l'Académie Impériale des Sciences de St. Pétersburg*, **17**, 257–261. Also *Oeuvres de P. L. Tchebychef*, (1899), **1**, (A. Markoff and N. Sonin, eds.). St. Petersburg, pp. 379–389. Reprinted (1961) by Chelsea, New York.

Tchebychef, P. L. (1864). Sur l'interpolation. Reprinted in *Oeuvres de P. L. Tchebychef*, (1899), **1**, (A. Markoff and N. Sonin, eds.) St. Petersburg, pp. 539–560. Reprinted (1961) by Chelsea, New York.

Tchebychef, P. L. (1875). Sur l'interpolation des valeurs equidistantes. Also *Oeuvres de P. L. Tchebychef*, (1899), **2**, (A. Markoff and N. Sonin, eds.) St. Petersburg, pp. 217–242. Reprinted (1961) by Chelsea, New York.

Terquem, O. (1838). Solution d'un probléme de combinaison. *Journal de Mathématiques Pures et Appliquées*, **3**, 559–560.
Tricomi, F. (1948). *Serie Ortogonali di Funzioni.* S. I. E. Istituto Editoriale Gheroni, Torino.
Vilenkin, N. Ja. and A. V. Klimyk (1991). *Representations of Lie Groups and Special Functions.* **1**, Kluwer, Dordrecht.
Weber, M. and A. Erdélyi (1952). On the finite difference analogue of Rodrigues' formula. *The American Mathematical Monthly*, **59**, 163–168.
Wilson, J. (1980). Some hypergeometric orthogonal polynomials. *SIAM Journal on Applied Mathematics*, **11**, 690–701.

CHAPTER 7

Olinde Rodrigues and Combinatorics

ULRICH TAMM

Introduction

After more than twenty years of absence from mathematics, Olinde Rodrigues returned to publishing mathematical papers in 1838, as discussed in Chapter 2, with four short notes on combinatorial problems (Rodrigues 1838*a*, 1838*b*, 1838*c*, 1839).

Rodrigues's contributions of 1838 were part of a discussion in the pages of the *Journal de Liouville* that Olry Terquem (1782–1862) had initiated. Euler had provided a recursion for the number of different ways in which a polygon may be partitioned into triangles, but he never published a proof (although he evidently had one), and neither did later authors manage to provide one. Terquem informed Liouville of this fact, and he later communicated the problem to several other geometers. It was Rodrigues who, by careful analysis of Lamé's first proof from 1838, gave a direct derivation of the Euler identity. He found an even simpler proof for the equivalent problem of determining the number of different parenthesizations of a nonassociative expression, which had been addressed shortly before by Catalan. We shall take a closer look at the discussion in the *Journal de Liouville* in order to emphasize Rodrigues's contributions and to demonstrate that Catalan's part in this subject was not that important.

Rodrigues worked on a second combinatorial problem in his 1839 paper, also first discussed in the *Journal de Liouville* by Terquem. Here, he gave the generating function for the enumeration of permutations with a given number of inversions. We shall also briefly discuss another result of Rodrigues relevant to combinatorial theory, which is contained in his later paper (Rodrigues 1843), where he derived an approximation for the central binomial coefficient $\binom{2n}{n}$. Furthermore, we shall mention an application of Rodrigues's formula for orthogonal polynomials in the enumeration of alternating-sign matrices. We include this topic, because this application was so important that Bressoud (Bressoud 1999) devoted a whole section to Rodrigues in his book.

Finally, this survey will conclude with some remarks on Rodrigues's way of approaching combinatorial problems, on the role of the book by Netto (Netto 1901) in the propagation of Rodrigues's results, and on the fact that the problems that he considered are still under discussion today.

Catalan numbers

The Catalan numbers are the solution to a variety of enumeration problems of which Stanley (Stanley 1999) lists more that seventy works. For $n = 0, 1, 2, \ldots$ the

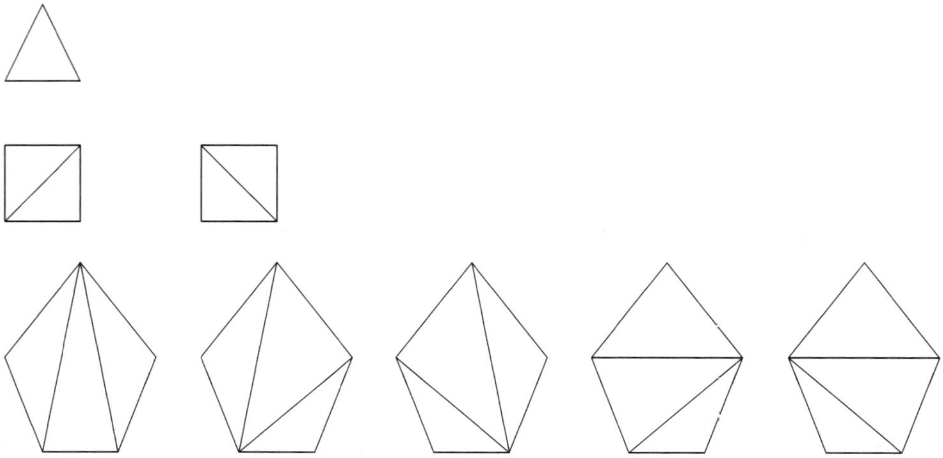

FIGURE 1. Partition of $(n+2)$-gons into triangles by nonintersecting diagonals.

nth *Catalan number* C_n is defined as

$$C_n = \frac{1}{n+1}\binom{2n}{n}. \tag{1}$$

Historically the first such problems were the polygon problem and the parenthesization problem, which were first discussed by Euler and Catalan, respectively.

Polygon Problem (Euler 1751): The Catalan number C_n, for $n = 1, 2, \ldots$, is the number of different partitions of an $(n+2)$-gon into triangles by means of nonintersecting diagonals (see Figure 1).

Parenthesization Problem (Catalan 1838): The Catalan number C_n, for $n = 1, 2, \ldots$, is the number of different parenthesizations of a nonassociative expression, where an expression in $n+1$ variables (in a fixed order) is evaluated by successive application of a binary operation $\circ$, e.g.,

$$\begin{aligned} n = 1 &: (a \circ b), \\ n = 2 &: ((a \circ b) \circ c), (a \circ (b \circ c)), \\ n = 3 &: (((a \circ b) \circ c) \circ d), ((a \circ (b \circ c)) \circ d), ((a \circ b) \circ (c \circ d)), \\ &\quad (a \circ ((b \circ c) \circ d)), (a \circ (b \circ (c \circ d))). \end{aligned}$$

Further problems, more important for current applications, in which the Catalan numbers occur as a counting function, have been considered in the nineteenth century by Cayley (Cayley 1859)—enumeration of binary, rooted trees—and by Whitworth (Whitworth 1879) and Bertrand (Bertrand 1878*a*)—enumeration of sequences under priority conditions, and the ballot problem.

Most of these problems can be extended in a natural way to enumerate similar configurations by the nth *generalized Catalan number* $C_n^{(m)}$, for a positive integer $m \geq 3$, defined as

$$C_n^{(m)} = \frac{1}{(m-1)n+1}\binom{mn}{n}. \tag{2}$$

One might, for instance, consider partitions of polygons into $(m+1)$-gons, expressions with m variables in one pair of parentheses, m-ary trees, etc.

The notion of Catalan numbers is standard nowadays in combinatorics. Catalan gave the explicit expression (1), which is implicit in Euler's recursion (3) below. In other branches of mathematics, they might also be known under different names. As Paulo Almeida pointed out to the author, rooted trees in quantum field theory are counted by CM-numbers (Connes-Moscovici). The notion of generalized Catalan numbers is not standard. Graham, Knuth, and Patashnik (Graham, Knuth, and Patashnik 1988, pp. 344–350) suggest that they be called Fuss numbers, but, perhaps, Binet numbers would be more appropriate, as we shall see later on.

The discussion in the *Journal de Liouville*: 1838–1843

For the early history of the Catalan numbers we refer to (Larcombe and Wilson 1998), where the presentation of Brown (Brown 1965) is corrected in some points. Seemingly, the problem of partitioning an n-gon into triangles was already discussed in China around 1730. Developments in Europe started in 1751 with a letter from Euler to Goldbach. Euler (Euler 1751) gave results for polygons of $3, \dots, 10$ sides using the formula $P_n = \frac{2 \cdot 6 \cdot 10 \cdots 2(2n-5)}{2 \cdot 3 \cdot 4 \cdots (n-1)}$ which is based on the recursion

$$P_{n+1} = \frac{4n-6}{n} P_n, \tag{3}$$

where P_n denotes the number of different partitions. He did not provide a derivation but stated that arriving at the proof had involved some considerable effort.

Euler also communicated the problem to von Segner, who produced a second recursion in (von Segner 1758):

$$P_{n+1} = P_n + P_3 P_{n-1} + P_4 P_{n-2} + \cdots + P_{n-2} P_4 + P_{n-1} P_3 + P_n. \tag{4}$$

Von Segner did not derive a formula for the numbers P_n from this recursion but only calculated the first values. A propagation error at P_{15} was, in turn, corrected by Euler (Euler 1758) on using (3). About thirty years later, Fuss (Fuss 1795), who received the problem from Pfaff, considered the more general problem of partitioning an n-gon into m-gons for a given m. He gave a recursion similar to von Segner's but did not find a formula for the numbers in question (in fact, these are just the generalized Catalan numbers).

This was state of the art in 1838, when the discussion in the *Journal de Liouville* began. We shall now go into this discussion more closely. In his biography of Liouville, Lützen (Lützen 1990, p. 102) states that '...Liouville did not just sit and wait for papers to be sent to the *Journal*. He often encouraged his colleagues and students to write papers, and he was often inspired to pursue ideas in the papers sent to the *Journal*.' For example, in 1839, after Liouville had told Jacques Binet about Rodrigues's work (Rodrigues 1838*a*) on the number of ways one can divide a polygon, Binet wrote a paper on the same subject (Binet 1839), which inspired Catalan (Catalan 1838) and Lamé (Lamé 1838) to similar studies, and a few years later Liouville (Liouville 1843) contributed to the discussion with a critical mention of a paper by Nicolas Fuss. In turn Binet (Binet 1843) commented on this in a note.

The problem was, in fact, brought to Liouville's attention by Terquem, who apparently had already found a solution, as stated in Liouville's footnote to Lamé (Lamé 1838): '...M. Terquem having succeeded with the help of certain properties of factorial has proposed this problem to me. I have consequently communicated it to several geometers, no one of them has solved it; M. Lamé has been more fortunate, I ignore if others had obtained before him such an elegant solution.' So it was Lamé (Lamé 1838) who published the first proof that the numbers obtained via Euler's recursion (3) may be derived from von Segner's recursion (4). His idea was to use a second recursion (the background is presented in the next section), namely

$$P_n = \frac{n}{2n-6}(P_3P_{n-1} + P_4P_{n-2} + \cdots + P_{n-2}P_4 + P_{n-1}P_3), \tag{5}$$

which then is simply plugged into (4) in order to obtain the recursion (3).

Catalan's first paper on the subject was published next to Lamé's article. It does not contain a proof but introduces the parenthesization problem and shows that it must have the same solution, because the solutions must obey the same recursion (4). Furthermore, Catalan derives the closed expression (1) from Euler's recursion (3), which is, of course, immediate.

In the same year 1838, Rodrigues published three small notes. In (Rodrigues 1838*a*), he gave a direct combinatorial proof (avoiding von Segner's recursion) for the Euler identity (4) via the polygon problem. This proof is based on Lamé's ideas. An even simpler proof, via the parenthesization problem, is presented in (Rodrigues 1838*b*), from which (4) can easily be derived. The third note, (Rodrigues 1838*c*), is a reflection on the binomial series, which comes into play in Binet's analytical derivation of (4). Binet (Binet 1839) obtained analytically the numbers (1) and hence the recursion (3) from von Segner's recursion (4). Note that $C_n = P_{n+2}$ for $n \geq 1$, so from von Segner's recursion (4) it is clear that the generating function

$$u(x) = C_0 + C_1x + C_2x^2 + \dots$$

satisfies the functional equation

$$u(x) = 1 + xu(x)^2.$$

This equation has the root $u(x) = \frac{1-\sqrt{1-4x}}{2x}$. The coefficients of the binomial series, applied to $(1-4x)^{1/2}$, finally yield the numbers (1).

Further papers on the subject were published in 1839 by Catalan (Catalan 1839*a*, 1839*b*) and by Duhamel (Duhamel 1839) in the *Journal de Liouville.* In the first issue of the journal *Archiv der Mathematik und Physik* its editor, Grunert (Grunert 1841), wanted to draw the attention of the above-mentioned French mathematicians to the paper of Fuss (Fuss 1795), who had considered the more general problem on the partition of an n-gon into m-gons: 'The first purpose of this paper is to draw the attention of the above-mentioned gentlemen to the fact, which seems completely missed by them, that already an older excellent mathematician Nicolaus Fuss, in *Novis Actis Academiae scientiarum imperialis Petropolitanae*, T. IX, p. 243 had made a start with a much more general problem presented to him, according to his own words, by Johann Friedrich Pfaff, namely the problem to determine the number of different ways in which an n-gon can be subdivided into m-gons by diagonals.' So Grunert suggested that the French mathematicians were totally unaware of the Fuss paper and that Fuss had made a start (ausgelöst) on a more general problem.

Two years later, in (Liouville 1843), this author made some remarks on Fuss's paper. This should be taken as a reply to Grunert, but Liouville did not mention his name at all. Perhaps the word 'ausgelöst' in the Fuss paper was interpreted as 'gelöst', which means solved, since Liouville pointed out quite sharply that Fuss had not solved the problem but in fact had missed a solution at several points. Liouville admitted that Fuss had already found a recursion, from which Lamé's recursion (5) would follow for the special case of the triangles, but he stated that Fuss had not compared his results with von Segner's recursion (4) and hence had missed Lamé's proof. Furthermore, Liouville wrote that Fuss had been aware of the functional equation

$$u(x) = 1 + xu(x)^{m-1}. \tag{6}$$

For $m = 3$, this equation was the central tool in Binet's analytical derivation of (3), which again had been missed by Fuss. Finally, Fuss did not realize either that even for the more general problem, a solution can be obtained from (6) by Lagrange inversion. This calculation was finally carried out by Binet (Binet 1843) in the note following that of Liouville (Liouville 1843). Binet derived the numbers in (2) via the Lagrange inversion as a solution to the problem discussed by Fuss. The Lagrange inversion turned out to be an important tool in algebraic enumeration (cf., e.g., (Gessel and Stanley 1996)).

Olinde Rodrigues's contributions to Catalan numbers

In order to understand the contribution to the subject given by (Rodrigues 1838*a*), we must briefly review the recursions (4) and (5).

Von Segner obtained his recursion (4) by using the following argument. Consider the polygon $(1, 2, 3, \ldots, n+1)$ on $n+1$ vertices, where sides of the polygon connect two vertices i and j, when $|i-j| = 1 \bmod n+1$. Now fix one side, $(1, n+1)$ say, and notice that the triangle $(1, n+1, k)$ subdivides the original polygon into two smaller polygons of size k and $n+2-k$, respectively, which can be partitioned into triangles in P_k and P_{n-k+2} ways. Since a triangle containing the side $(1, n+1)$ must be contained in every partition, on repeating this subdivision with all triangles $(1, n+1, k)$, $k = 2, 3, \ldots, n$, all partitions of the original polygon are obtained. Lamé and Fuss's idea behind the second recursion (5) is very similar: They fix a vertex, 1 say, and consider all diagonals $(1, k)$ that subdivide the original polygon $(1, 2, \ldots, n)$ on n vertices into two smaller polygons of size k and size $n-k+2$. Since not all partitions can be obtained this way, the procedure is repeated by fixing subsequently the vertices $i = 2, 3, \ldots, n$ and then subdividing the polygon with all diagonals (i, k). In this way, the polygon $(1, 2, \ldots, n)$ is partitioned into triangles n times and each triangle occurs in exactly $2n-6$ of the partitions, yielding the recursion (5).

Rodrigues (Rodrigues 1838*a*) observed that these two ideas may be combined to obtain a direct derivation of the recursion (3). He simply identified the two vertices 1 and $n+1$ of the $(n+1)$-gon in the first construction to obtain a single vertex, 1 say, in an n-gon. Of course, accordingly, the triangles collapse into diagonals. Careful enumeration gives the recursion (3) without using (4) and (5).

In his second paper on the subject, Rodrigues (Rodrigues 1838*b*) provided a second, even shorter, proof of (3) via the parenthesization problem. Notice that in the original problem the variables in the expression are in fixed order: $a_1 \circ a_2 \circ \cdots \circ a_n$, say. Rodrigues in fact derived a recursion for the total number B_n of all parenthesizations of expressions, where the variables $a_1, \ldots, a_n$ may occur in any order. Since there are $n!$ many permutations of the set of variables, finally one has to divide by $n!$ to obtain the Catalan numbers. Now assume that a new variable

a_{n+1} is inserted into an expression E, with a well-formed parenthesization of the variables $a_1, \ldots, a_n$. Either a_{n+1} is placed at the beginning or at the end, yielding the two expressions $(a_{n+1} \circ (E))$ and $((E) \circ a_{n+1})$, or a_{n+1} is inserted into one of the $n-1$ pairs of parentheses, which can be done in four ways for each such pair. To see this, assume that the parentheses combine the expressions C and D in E. Then the four possibilities for inserting a_{n+1} into $(C \circ D)$ are

$$((a_{n+1} \circ C) \circ D), ((C \circ a_{n+1}) \circ D), (C \circ (a_{n+1} \circ D)), (C \circ (D \circ a_{n+1})).$$

So, $B_{n+1} = 2B_n + 4(n-1)B_n = (4n-2)B_n$. By the previous considerations $P_{n+1} = \frac{B_n}{n!}$. The proof can be presented in an even shorter, different way, which Olinde Rodrigues probably had in mind. Each expression with n variables (in arbitrary order) may be regarded as a string of $4n-3$ symbols, namely the n variables $a_1, \ldots, a_n$, $n-1$ operators $\circ$, and $2n-2$ parentheses. Now, with a new variable a_{n+1}, a new operator $\circ$, and a new pair of parentheses must also be inserted. For the position of a_{n+1} there are exactly $4n-2$ choices, before or behind the previous expression or between any two symbols in the previous expression. The positions of the other new symbols $\circ$, (, and) to be inserted are then automatically determined.

In his third note from the same year, Rodrigues (Rodrigues 1838*c*) provided an algebraic and elementary derivation—without using the Taylor series—for the expansion of the binomial series $(1+x)^\alpha = \sum_{j=0}^{\infty} \binom{\alpha}{j} x^j$ when α is a negative integer or a rational number. This expansion for $\alpha = \frac{1}{2}$ was the central tool in Binet's analytical proof of Euler's recursion (3).

Netto's presentation

Netto (Netto 1901) devoted considerable space in his book to Catalan numbers without referring to them as such—(Netto, 1901, Sections 122–125) this author devoted Sections 122–125, Chapter 9, to Catalan numbers (but he did not use this notion yet). In Section 122 he first introduced the parenthesization problem, correctly stating in a footnote that it had first been treated by Catalan. He continued with Binet's analytical derivation of the Euler recursion (without mentioning Binet or Euler), and in Section 123 he gave Rodrigues's direct proof (referring to Rodrigues). In Section 124 he pointed out the connection to the problem of partitioning a polygon into triangles. His formulation, however, is extremely unhelpful: 'E. Catalan pointed out that the problem under discussion is connected to the geometrical question: in how many ways can one subdivide a polygon in the plane by diagonals?' To a reader, not familiar with the complete story, his presentation may suggest that, chronologically, the parenthesization problem was discussed before the polygon problem—especially since no further reference to Euler or Lamé (whose paper came before Catalan's) is given—and that everything started with Catalan. It is not clear why Netto presented the topic in this way. Perhaps the parenthesization problem was just more suitable for him because the rest of the chapter is devoted to similar enumeration problems due to Schröder or perhaps because he wanted to include a direct derivation and preferred Rodrigues's proof of the parenthesization problem to the one for the polygon partitioning problem. In any case his book was actually the first textbook on combinatorics (a previous book by Hindenburg published around 1800 did not circulate widely), and his presentation might have been the source from which misconceptions have originated.

Sequences and inversions

In his fourth paper, (Rodrigues 1839), published a year later than the other three, Rodrigues discussed the problem of counting the number of permutations of n elements with exactly i inversions. Indicating this number by $Z_{n,i}$, he showed that $Z_{n,i}$ is the coefficient of t^i in the generating function

$$(1+t)(1+t+t^2)(1+t+t^2+t^3)\cdots(1+t+t^2+\cdots+t^{n-1}) = \frac{\prod_{j=1}^{n}(1-t^j)}{(1-t)^n}.$$

On taking the derivative of this generating function, he obtained a new solution for a problem posed by Stern and later proved by Terquem, namely how to determine the sum of all the inversions of the $n!$ permutations on $\{1,2,\ldots,n\}$, which is $\frac{n!}{2}\cdot\frac{n(n-1)}{2}$. Finally, he further simplified the analysis of the numbers $Z_{n,i}$ by demonstrating that they can be written as

$$Z_{n,i} = \binom{n+i-1}{i} + \binom{n+i-2}{i-1}E_{1,1} + \binom{n+i-3}{i-2}E_{2,2} + \cdots + E_{i,i},$$

where $E_{n,i}$ is the coefficient of t^i in the product $\prod_{j=1}^{n}(1-t^j)$.

Netto (Netto 1901) included a chapter on sequences and inversions in his book without mentioning Rodrigues's results. Carlitz (Carlitz 1970), who derived the generating function for the more general problem of enumerating sequences with k_1 1's, k_2 2's , ..., k_n n's according to a given number of inversions, redressed this failure, C. A. Church having informed him of Rodrigues's work. Of course, for the special case $k_1 = k_2 = \cdots = k_n = 1$, the permutations discussed by Rodrigues arise. For further applications, Chapter 6 should be consulted.

Approximation of central binomial coefficients

It is important for many applications to have a good approximation of binomial or even multinomial coefficients. For instance, the entropy function of thermodynamics and information theory, defined as $H(p_1,\ldots,p_m) = -\sum_{i=1}^{m} p_i \cdot \log p_i$, for $p_1,\ldots,p_m \geq 0$ with $\sum_{i=1}^{m} p_i = 1$, may be obtained by approximation of the multinomial coefficients $\binom{n}{(np_1)\cdots(np_m)} \approx 2^{n\cdot H(p_1,\ldots,p_m)}$. Furthermore, the limit theorem of De Moivre and Laplace is derived by an appropriate approximation of the binomial coefficients. Usually, Stirling's formula is applied to the single factorials.

For various applications in probability theory an approximation of the central binomial coefficients $\binom{2n}{n}$ is of special importance. This was extensively discussed by Laplace (Laplace 1814) in his *Théorie Analytique des Probabilités*. A later application is the demonstration of the nonrecurrence of the random walk on the line, where the Catalan numbers come into play in the formula for the expected value.

Rodrigues (Rodrigues 1843) provides a new method for the approximation of $\binom{2n}{n}$ based on the development of trigonometric functions and on the Wallis product $\frac{\pi}{2} = \prod_{n=1}^{\infty}\frac{(2n)^2}{(2n)^2-1}$. As he points out, his method yields an approximation of the same order as the one resulting from the first three terms of the approximation presented by Laplace (who used the coefficients occurring in Stirling's formula) and additionally gives bounds for the error involved.

The Rodrigues formula and alternating-sign matrices

The last work of Olinde Rodrigues concerned with combinatorics that we shall discuss in this paper is not an original result of his, but it is an application of the formula for orthogonal polynomials still known by his name today. However, we include it here since it is quite unusual to find a whole section of a mathematics book devoted to Olinde Rodrigues, as is the case in the book by Bressoud (Bressoud 1999) on alternating-sign matrices. An alternating-sign matrix is a square matrix with entries from $\{0, 1, -1\}$ such that (i) the entries in each row and column sum up to 1 and (ii) the nonzero entries in each row and column alternate in sign. An example is

$$\begin{pmatrix} 0 & 0 & 1 & 0 & 0 \\ 1 & 0 & -1 & 0 & 1 \\ 0 & 0 & 1 & 0 & 0 \\ 0 & 1 & -1 & 1 & 0 \\ 0 & 0 & 1 & 0 & 0 \end{pmatrix}.$$

These matrices were discovered in the analysis of an algorithm to evaluate determinants invented by Charles Dodgson, better known as Lewis Carroll. There are also deep relations with problems in statistical mechanics, such as the configuration of water molecules in 'square ice', which can be described by an alternating-sign matrix.

The alternating-sign matrix conjecture concerns the total number of $n \times n$ alternating-sign matrices, which was conjectured to be $\prod_{j=0}^{n-1} \frac{(3j+1)!}{(n+j)!}$. The problem was open for fifteen years until it was finally settled by Zeilberger (Zeilberger 1996). A discrete version of the Legendre polynomials comes in as an important step in the proof. Bressoud (Bressoud 1999, p. 252) points out in his section on Rodrigues that '[t]his work would be naught unless we had an alternate description of the polynomials $P_n(x)$. In 1816 Olinde Rodrigues observed that the nth Legendre Polynomial could be written as $\frac{n!}{(2n)!} D^n[(x^2-1)^n]$.'

Mathematicians are still looking for a simpler and more natural derivation of the formula for the total number of alternating-sign matrices. This formula also arises as a Hankel determinant where the entries in the matrix are the coefficients of the generating function $\frac{1-(1-9x)^{1/3}}{3x}$, the binomial series studied by Rodrigues (Rodrigues 1838*c*) with exponent $\frac{1}{3}$. An appropriate combinatorial interpretation of these numbers might yield a new proof of the alternating-sign matrix conjecture. An idea would be to look for an explanation of the recursion $P_{n+1} = \frac{6n-4}{n} P_n$ in the spirit of Rodrigues (Rodrigues 1838*b*) and then to find a bijection to a problem concerning disjoint lattice paths, which are enumerated by Hankel determinants.

Concluding remarks

Olinde Rodrigues's approach to combinatorial problems led to very elegant solutions. This holds for the direct proofs for the polygon and parenthesization problems as well as for the derivation of the generating function for the numbers of permutations according to their inversions. His constant search for simplicity in his proofs is striking. Some quotations from his papers may support this impression: '...formula that M. Laméhas proved in the first issue of this Journal and that can be established more directly below' (Rodrigues 1838*a*, p. 547). It is clear that

Rodrigues looks whenever it is possible for elementary proofs: '[I] give here the direct and elementary solution' (Rodrigues 1838*b*, p. 549); '[e]lementary and purely algebraic proof...' (Rodrigues 1838*c*). Also, simplicity is never far from his mind: '[o]ne can, however, arrive, more simply at this result' (Rodrigues 1839, p. 238) or '[t]his is the simplest expression which I have found' (Rodrigues 1839, p. 239).

The two combinatorial problems on which Rodrigues worked in 1838 and 1839, were brought to the interest of French mathematicians by Olry Terquem, although little is known at present about the connection between Rodrigues and Terquem.

Contrasting with almost all his other results, Rodrigues's work on the Catalan numbers is still remembered nowadays. As mentioned before, Netto (Netto 1901) included the proof from (Rodrigues 1838*b*) in his book, and Rodrigues is also frequently cited in further publications on this topic. Strangely, Netto (Netto 1901) does not refer to Rodrigues's paper of 1839, although in the same book he also addresses the problem of enumerating permutations with a given number of inversions. Rodrigues's result on the generating function for these permutations was forgotten until Carlitz (Carlitz 1970) referred to it. Even nowadays it is not widely known that this generating function goes back to Rodrigues; in a recent paper on the subject Margolius ((Margolius 2001), for instance) attributed it to a work by Muir of 1898.

The reference of Margolius also demonstrates that the topics that Olinde Rodrigues considered are still under discussion today. This is even more obvious for the Catalan numbers, which are quite in fashion at the present time, their role in the enumeration of binary trees having made them of importance in computer science. Such trees may serve as a data structure, for instance in data compression (cf. (Kobayashi, Morita, and Hoshi 1996)). Knuth (Knuth 1973) addressed the problem of sorting a sequence through a stack, which is another application of the Catalan numbers. Generalizations to sorting problems with two or more stacks have recently been discussed so it is not surprising that some new books on discrete mathematics devote a chapter to Catalan numbers. For new combinatorial developments in this context the survey by Aigner (Aigner 2001) should be consulted, and their relation to orthogonal polynomials is also discussed in (Tamm 2001). In this way, Hankel matrices of generalized Catalan numbers are related to alternating-sign matrices. There is no better way to gauge the originality and importance of Rodrigues's combinatorial works than to quote a comment on Rodrigues's mathematical talent by Bertrand(Bertrand 1878*b*). 'A very elementary problem, apparently very simple, made it soon evident the flexibility of his talent. Segner had proposed it in the last century, and two very different formulae—the most elegant due to Euler—had necessarily to provide identical results; but this identity was not easy to establish [....] Olry Terquem has succeeded by a long and tortuous route; he proposed the problem to M. Liouville who communicated it to several geometers none of whom succeeded in solving it. Lamé was luckier and sent a solution the day after he had received the enunciation. The first place in this fight [...] was not, however, reserved for him; Lamé's elegant solution woke up an eminent spirit, for a long time oblivious to science. Olinde Rodrigues, in a note a few pages long, reminded his old fellow students, who by then had become great masters, that in the past, at the *lycée* he led them all and that, had he wished, he could have become their equal.'

Bibliography

Aigner, M. (2001). Catalan and other numbers: a recurrent theme. In *Algebraic combinatorics and computer science. A tribute to Gian-Carlo Rota* (H. Crapo and D. Senato, eds.), pp. 347–390. Springer-Verlag Italia, Milan.

Bertrand, J. (1878*a*). Solution d'un problème. *Comptes Rendus de l'Académie des Sciences*, Paris, **105**, 369.

Bertrand, J. (1878*b*). Élogue de Gabriel Lamé, ingénieur en chef the mines lu dans la scéance publique annuelle du 28 janvier 1878. *Annales de Mines*, **7**, 13.

Binet, J. (1839). Réflexions sur le problème de déterminer le nombre de manières dont une figure rectiligne peut être partagée en triangles au moyen de ses diagonales. *Journal de Mathématiques Pures et Appliquées*, **4**, 79–91.

Binet, J. (1843). Note de M. J. Binet. *Journal de Mathématiques Pures et Appliquées*, **8**, 394–396.

Bressoud, D. M. (1999). *Proofs and confirmations: The story of the alternating sign matrix conjecture*. Cambridge University Press.

Brown, W. G. (1965). Historical note on a recurrent combinatorial problem. *American Mathematical Monthly*, **72**, 973–977.

Carlitz, L. (1970). Sequences and inversions. *Duke Mathematical Journal*, **37**, 193–198.

Catalan, E. (1838). Note sur une équation aux différences finies. *Journal de Mathématiques Pures et Appliquées*, **3**, 508–516.

Catalan, E. (1839*a*). Solution nouvelle de cette question: un polygone étant donné, de combien de manières peut-on le partager en triangles au moyen de diagonales. *Journal de Mathématiques Pures et Appliquées*, **4**, 91–94.

Catalan, E. (1839*b*). Addition à la note sur une équation aux différences finies insérée dans le volume précédent, page 508. *Journal de Mathématiques Pures et Appliquées*, **4**, 95–99.

Cayley, A. (1859). On the analytical forms called trees—part II. *Philosophical Magazine, 4th series*, **18**, 374–378.

Duhamel, J. M. C. (1839). Intégration d'une équation aux différences. *Journal de Mathématiques Pures et Appliquées*, **4**, 222–224.

Euler, L. (1751). Letter to Christian Goldbach dated 4 September 1751.

Euler, L. (1758). Enumeratio modorum, quibus figurae planae rectilineae per diagonales dividintur in triangula, auctore J. A. de Segner, page 203. *Novi Commentarii Academiae Scientiarum Imperialis Petropolitanae*, **7**, 13–15.

Fuss, N. (1795). Solutio quaestionis, quot modis polygonum n laterur in polygonam laterum, per diagonales resolvi queat. *Nova Acta Academiae Scientiarum Imperialis Petropolitanae*, **9**, 243–251.

Gessel, I. and Stanley, R. P. (1996). Algebraic enumeration. In *Handbook of combinatorics, Vol. 2* (R. L. Graham, M. Grötschel, and L. Lovasz, eds.), pp. 1021–1069, Elsevier Science B. V., Amsterdam.

Graham, R. L., D. E. Knuth, and O. Patashnik (1988). *Concrete mathematics*. Addison Wesley, Reading.

Grunert, J. A. (1841). Über die Bestimmung der Anzahl der verschiedenen Arten, auf welche sich ein $n$eck durch Diagonalen in lauter $m$ecke zerlegen lässt, mit Bezug auf einige Abhandlungen der Herren Lamé, Rodrigues, Binet, Catalan und Duhamel in dem Journal de Mathématiques pures et appliquées, publié par Joseph Liouville, Vols. 3, 4. *Archiv der Mathematik und Physik*, **1**, 193–203.

Knuth, D. E. (1973). *The art of computer programming*, Vol. 1, *Fundamental algorithms*. Addison Wesley, Reading.

Kobayashi, K., Morita, H., and Hoshi, M. (1996). Enumerative coding for k-ary trees. *Proceedings of the 19th Symposium on Information Theory and its Applications (SITA96)*, Hakone, Japan, pp. 377–379.

Lamé, G. (1838). Extrait d'une lettre de M. Lamé à M. Liouville sur cette question: un polygone convexe étant donné, de combien de manières peut-on le partager en triangles au moyen de diagonales? *Journal de Mathématiques Pures et Appliquées*, **3**, 505–507.

Laplace, P. S. (1814) *Théorie analytique des probabilités (2nd ed., rev. et augm. par l'auteur)*. Courcier, Paris.

Larcombe, P. J. and Wilson, P. D. C. (1998). On the trail of the Catalan sequence. *Mathematics Today*, **34**, 114–117. (Also *ibid*, **35**, 25 and 89).

Liouville, J. (1843). Remarques sur un mémoire de N. Fuss. *Journal de Mathématiques Pures et Appliquées*, **8**, 391–394.

Lützen, J. (1990). *Joseph Liouville 1809–1882: Master of pure and applied mathematics.* Studies in the History of Mathematics and Physical Sciences, Vol. 15. Springer-Verlag, New York.

Margolius, B. H. (2001). Permutations with inversions. *Journal of Integer Sequences*, **4**, Article 01.2.4, 13 pp. (electronic).

Netto, E. (1901). *Lehrbuch der Combinatorik.* Teubner, Leipzig.

Rodrigues, O. (1838*a*). Sur le nombre de manières de décomposer un polygone en triangles au moyen de diagonales. *Journal de Mathématiques Pures et Appliquées*, **4**, 547–548.

Rodrigues, O. (1838*b*). Sur le nombre de manières d'effectuer un produit de n facteurs. *Journal de Mathématiques Pures et Appliquées*, **4**, 549.

Rodrigues, O. (1838*c*). Démonstration élémentaire et purement algébrique du développement d'un binôme élevé à une puissance négative ou fractionnaire. *Journal de Mathématiques Pures et Appliquées*, **4**, 550–551.

Rodrigues, O. (1839). Note sur les inversions, ou dérangements produits dans les permutations. *Journal de Mathématiques Pures et Appliquées*, **5**, 236–240.

Rodrigues, O. (1843). Du développement des fonctions trigonométriques en produits de facteurs binômes. *Journal de Mathématiques Pures et Appliquées*, **8**, 217–224.

Stanley, R. P. (1999). *Enumerative combinatorics*, Vol. 2. Cambridge University Press.

von Segner, J. A. (1758). Enumeratio modorum, quibus figurae planae rectilineae per diagonales dividintur in triangula. *Novi Commentarii Academiae Scientiarum Imperialis Petropolitanae*, **7**, 203–210.

Tamm, U. (2001). Some aspects of Hankel matrices in combinatorics and coding theory. *The Electronic Journal of Combinatorics*, **8**, # A1, 31 pp. (electronic).

Whitworth, M. A. (1879). Arrangements of m things of one sort and n things of another sort, under certain conditions of priority. *Messenger of Mathematics*, **8**, 105–114.

Zeilberger, D. (1996). Proof of the refined alternating sign matrix conjecture. *New York Journal of Mathematics*, **2**, 59–68.

CHAPTER 8

Olinde Rodrigues's Paper of 1840 on a Group of Transformations

Jeremy Gray

Introduction

In this chapter we discuss a paper by Olinde Rodrigues which became almost completely forgotten and which is perhaps the first treatment of the subject of groups of motions. This is his '*Des lois géométriques qui régissent les déplacements d'un systéme solide dans l'espace, et de la variation des coordonnées provenant de ces déplacements considérés indépendamment des causes qui peuvent les produire*', which was published in 1840 in Liouville's *Journal de Mathématiques*, Volume 5, pages 380–440. We shall proceed by discussing (1) the contents, (2) the context, and (3) the significance of the paper.[1] The circumstances of its publication in relation to Rodrigues's life are discussed in Chapter 2.

The contents of Rodrigues's paper

As the title of his paper indicates, Rodrigues studied the motions (déplacements) of a rigid body (système solide) in three dimensional space independently of any dynamical considerations. He began by giving a complete description of motions in synthetic terms, establishing successively that a body is fixed in space once three noncollinear points have been determined; that if two points are fixed the motion is a rotation about an axis through those points; that translations when composed give a translation which is independent of the order of composition and can be found by the parallelogram law for translations ('loi du polygone des translations' (Rodrigues 1840, p. 383)); and that a translation is equal to an infinitesimal rotation about an axis perpendicular to the direction of the translation but situated at an infinite distance.[2] This last observation allowed him to consider translations as a special class of infinitesimal rotations.

To describe a general motion, Rodrigues noticed, as Euler ((Euler 1758, §2), (Euler 1765, §690)) had before him, that it can always be factored as a rotation followed by a translation. Indeed, let P be any point on the solid and P' its image under the motion. The translation taking P to P' puts the solid in what Rodrigues called an intermediate position. The rotation about P which moves the solid from its initial position to its intermediate position, followed by the translation from

[1]This is a modified version of a paper published in *Archive for History of Exact Sciences*, **21** (1980), 375-385. Reprinted with kind permission of Springer Science and Business Media. Material covered elsewhere in this book has been removed, and the references have been updated.

[2]Rodrigues described the rotation as 'of an amplitude infinitesimally small around a fixed axis infinitely distant and normal to the direction of this translation' (Rodrigues 1840, p. 381).

P to P', is the sought-for factorization of the original motion. This factorization is not unique, for the position of the axis of rotation may be altered if a different point P is chosen. However, he showed that this only varies the position of the axis, not its direction or amount (Rodrigues 1840, p. 385). This result follows from the fact that a rotation about one axis is equal to a rotation through an equal amount about a parallel axis, followed by the translation which sends a point on the second axis to its image under the first rotation, and this observation seems to be due to Rodrigues. The nature of the factorization permitted him to define the absolute translation, t, of any motion (Rodrigues 1840, p. 386) as the value of the projection of PP' onto the axis of the rotation part of the motion, where by the projection of PP' he meant the feet of the perpendiculars from P and P' to the axis. This value must be a constant independent of P, and it vanishes precisely for those motions which are composed of a rotation and a translation perpendicular to the axis of rotation, which are, of course, pure rotations about a suitably chosen parallel axis. The absolute translation is the measure of the translation part of screwlike motions, those composed of a rotation and a translation parallel to the axis of rotation, and the position of the axis of a general motion can always be chosen so that it is the resultant (résultant) of a rotation and a translation along the axis (Rodrigues 1840, *Théorème fondamental*, p. 385).

Rodrigues showed that any motion can be factorized in infinitely many ways as a product of two rotations about fixed intersecting (convergents) axes and then concentrated on theorems about the composition of motions. He had already considered the composition of two translations and of a translation and a rotation. He had finally to consider the composition of two rotations about axes which might be coincident or parallel or nonintersecting. The case of nonintersecting axes reduces to that of intersecting axes once the composition of a rotation and a translation is understood, for the axes can be made to coincide by means of a translation. In the case of parallel axes the combined motion is a rotation about a third, parallel, axis, the position of which depends on the order of composition. The combined effect of two rotations about coincident axes, he showed, was again a rotation, the size of which did not depend on the order of composition but the axis of which did depend essentially upon the order. However, he remarked, the combination of two infinitesimal rotations is independent of their order, for they combine like infinitesimal translations which can be computed by resolving them as vectors along coordinate axes.

The synthetic description of motions in space being now concluded, Rodrigues then reexpressed all his results in analytic terms. Of most interest to us is his derivation of the equations corresponding to a rotation expressed in terms of four parameters and his discovery of the rule for combining these parameters when two rotations are performed successively.

To describe a rotation, Rodrigues took a rectangular coordinate system Ox, Oy, Oz with origin on the axis of rotation, and he specified the axis in terms of its direction cosines $\cos g, \cos h, \cos l$ with respect to the coordinate axes; he called the amplitude of the rotation θ .

The formulae for a general motion involving a rotation and a translation were not expressed by Rodrigues in a particularly simple form (Rodrigues 1840, p. 399), but for a pure rotation they reduce to a more elegant set as follows. Suppose the point (x, y, z) goes to $(x + \Delta' x, y + \Delta' y, z + \Delta' z)$. Let (X', Y', Z') be the midpoint of the motion, i.e., $(X', Y', Z') = (x+\Delta' x/2, y+\Delta' y/2, z+\Delta' z/2)$; then he showed (Rodrigues 1840, p. 407, correcting misprints):

$$\Delta' x = 2\tan\left(\frac{\theta}{2}\right)(Y'\cos l - Z'\cos h),$$
$$\Delta' y = 2\tan\left(\frac{\theta}{2}\right)(Z'\cos g - X'\cos l),$$
$$\Delta' z = 2\tan\left(\frac{\theta}{2}\right)(X'\cos h - Y'\cos g)$$

from which the nowadays more familiar expressions relating $(x + \Delta' x, y + \Delta' y, z + \Delta' z)$ to (x, y, z) may easily be obtained, although Rodrigues did not do so.

His next task was to follow the first rotation with a second, parameterized by θ', g', h', l', and to express the parameters Θ, G, H, and L of the combined rotation in terms of the eight parameters $\theta, g, \ldots, l'$. In the simplest case the two axes of rotation meet at the coordinate origin. He then found by elementary algebra that if v is the angle between the two axes, so

$$\cos v = \cos g \cos g' + \cos h \cos h' + \cos l \cos l',$$

then

$$\tan\left(\frac{\Theta}{2}\right)\cos G = \frac{\tan(\frac{\theta}{2})\cos g + \tan(\frac{\theta'}{2})\cos g' + \tan(\frac{\theta}{2})\tan(\frac{\theta'}{2})(\cos h \cos l' - \cos l \cos h')}{1 - \tan(\frac{\theta}{2})\tan(\frac{\theta'}{2})\cos v}$$

with similar formulae for $\tan(\frac{\Theta}{2})\cos H$ and $\tan(\frac{\Theta}{2})\cos L$, and for the amplitude he found $\cos\frac{\Theta}{2} = \cos\frac{\theta}{2}\cos\frac{\theta'}{2} - \sin\frac{\theta}{2}\sin\frac{\theta'}{2}\cos \nu$.

For the inclination of the axis of the combined rotation he found

$$\sin\frac{\Theta}{2}\cos G = \sin\frac{\theta}{2}\cos\frac{\theta'}{2}\cos g + \sin\frac{\theta'}{2}\cos\frac{\theta}{2}\cos g' + \sin\frac{\theta}{2}\sin\frac{\theta}{2}(\cos h \cos l' - \cos l \cos h')$$

with similar formulae for $\sin\frac{\Theta}{2}\cos H$ and $\sin\frac{\Theta}{2}\cos L$.

He remarked that Θ is clearly independent of the order of combination of the two rotations. However, if one sets $\cos l = 0 = \cos l', \cos h = 0, \cos g = 1, \cos g' = \cos v, \cos h' = \sin v$, as one may without loss of generality, then $\sin(\frac{\Theta}{2})\cos L$ reduces to $\sin(\frac{\theta}{2})\sin(\frac{\theta'}{2})\sin v$ which changes sign when the order of combination is reversed, whereas $\sin(\frac{\Theta}{2})\cos G$ and $\sin(\frac{\Theta}{2})\cos H$ are unaltered. In fact, this is only correct as far as the expression for $\sin(\frac{\Theta}{2})\cos G$ is concerned, but that did not affect the validity of his conclusion, which was that the order of combining two rotations affects the axis of the final rotation but not its amplitude, as he had earlier shown synthetically.

Rodrigues gave analytic formulae for all the possible motions and their combinations, finite and infinitesimal, rotations about various axes, and translations. The infinitesimal rotation is of some interest. He found that it was described by the equations (Rodrigues 1840, p. 402)

$$\delta x = \alpha + \theta y \cos l - \theta z \cos h,$$
$$\delta y = \beta + \theta z \cos g - \theta x \cos l,$$
$$\delta z = \gamma + \theta x \cos h - \theta y \cos g$$

about a central axis defined by the equations

$$\frac{\theta x + \gamma \cos h - \beta \cos l}{\cos g} = \frac{\theta y + \alpha \cos l - \gamma \cos g}{\cos h} = \frac{\theta z + \beta \cos g - \alpha \cos h}{\cos l}.$$

He then showed that two infinitesimal rotations about the same axis combined additively (Rodrigues 1840, p. 414). On pp. 421–430 he rederived his equations for a general motion purely algebraically, and then on pp. 430–432 he made the significant observation that the infinitesimal motions can be integrated to give the finite motions.[3] Specifically, letting the finite change in x be Δx and the infinitesimal change be δx, etc., he deduced from $\Delta(dx^2 + dy^2 + dz^2) = 0$, which expresses the rigidity of the body, that the following six equations for (x, y, z) and $(X, Y, Z) = (x + \frac{1}{2}\Delta x, y + \frac{1}{2}\Delta y, z + \frac{1}{2}\Delta z)$ hold (Rodrigues 1840 p. 431, correcting misprints):

$$\frac{d}{dX}\Delta x = 0 = \frac{d}{dY}\Delta y = \frac{d}{dZ}\Delta z,$$
$$\frac{d\Delta x}{dY} + \frac{d\Delta y}{dX} = 0 = \frac{d\Delta x}{dZ} + \frac{d\Delta z}{dX} = \frac{d\Delta y}{dZ} + \frac{d\Delta z}{dY}.$$

These integrate easily to give $\Delta x = A + pY - nZ, \Delta y = B + mZ - pX, \Delta z = C + nX - mY$ from which his earlier formulae for a general motion may be obtained. The paper concludes with a brief application of the infinitesimal analysis to a problem in statics. He establishes what he calls the equation of virtual velocities ('équation des vitesses virtuelles' (Rodrigues 1840, pp. 438–439) which says: 'the forces acting on a solid system being in equilibrium, if the system is moved infinitely little from its actual position by whatever cause, the sum of the forces multiplied by the infinitely small distances the points of the system run through in the direction of these forces must be zero, and conversely, and this is the statement of the principle of virtual velocities.'

The context of the paper

To appreciate the significance of Rodrigues's paper, it is necessary to look briefly at what work on the subject of rigid body motions had already been done. That subject forms part of dynamics and statics, an immense field of mathematical research, but it is precisely mechanics from which Rodrigues was seeking to abstract his discoveries. By studying the laws governing the displacements of a rigid body independently of the causes which can produce them, he made an abrupt break with the tradition of Euler. In Euler's work it is the forces upon a rigid body and their effects which are discussed, and the geometry of motions is developed only incidentally. Thus Euler did establish that every motion can be expressed as a combination of a rotation and a translation (Euler 1758, 1765) and that every rotation of a rigid body is about an axis which is instantaneously fixed (Euler 1758, §3). But his chief concern was with the effect of impressing given forces upon

[3]Formulae for infinitesimal motion were given by Lagrange in his *Mécanique Analitique* of 1788 in the section 'La Statique', pp. 55–58, where the composition was stated to be additive.

a body and in finding what forces will produce a desired effect. The techniques Euler employed are those of analysis. In Rodrigues's paper all that is considered are the abstract motions. So it seems that Euler is the first to give an explicit notation for describing a rotation, the so-called 'Euler' angles, but that Rodrigues is the first to give a formula for the combination of two rotations as expressed in terms of four parameters, as Klein remarked (Klein 1884, p. 38). However, Jacobi observed (Jacobi 1881) that Euler had considered the combination of orthogonal matrices[4] in his number-theoretic research (Euler 1770), and there he composed the general orthogonal rotation in three, four, or five dimensions as a product of rotations in successive planes, as he also did in his mechanics (Euler 1775*a*) in three dimensions. Formulae for finite motions were given by Monge (Monge 1784) and in a simplified form by Chasles (Chasles 1837*b*, p. 678), neither of whom sought the rule for combining motions. Chasles' geometrical derivation (Chasles 1830) of Euler's results in $\mathbb{R}^3$ influenced Rodrigues's methods, and Rodrigues attributed some of the theorems about a single motion to Chasles (Rodrigues 1840, p. 386).

The aim of Rodrigues's paper, though, was not to calculate the composition of two rotations, a rather easy and uninteresting task, but to draw attention to the motions of a rigid body as an important object of study. Euler gave a set of four parameters which describe a rotation: p, q, r, and $\cos\phi$ which satisfy the equation $p^2 + q^2 + r^2 = 1$ for a fixed but arbitrary ϕ (Euler 1775*b*, p. 124). Rodrigues supplied a similar set of parameters g, h, l, and θ and gives the formulae for the composition of two rotations in terms of these parameters. As Klein (op. cit.) remarked, this is equivalent to Hamilton's system of quaternions. Klein incorrectly implied that geometric research governed Hamilton's research prior to his discovery of quaternions. In fact Hamilton came to his discovery of quaternions at the end of a long series of algebraic research, and he said that he only discovered the connection between unit quaternions and rotations of the unit sphere the next day (Hamilton 1843*a*, 1843*b*). Klein here was guided by Cayley's remarks (Cayley 1845) in which Cayley observed that if q denotes the quaternion $\alpha + \lambda i + \mu j + \nu k$ and $xi + yj + zk$ is a unit quaternion (and so lies on the unit sphere in $\mathbb{R}^3$), then a rotation of the sphere is given by $(ix + jy + kz) \to q^{-1}(ix + jy + kz)q$. This rotation[5] is through θ about an axis through the centre of the sphere which has direction cosines $\cos f, \cos g, \cos h$ given by $\lambda = \tan(\frac{1}{2}\theta) - \cos f, \mu = \tan(\frac{1}{2}\theta) - \cos g, \nu = \tan(\frac{1}{2}\theta) - \cos h$. This is the first published account of the connection between quaternions and rotations, but it is not the first time it was discovered (cf. (Gauss 1819)).[6] Cayley conceded priority in a footnote to the paper when it was republished in the first volume of his *Collected Mathematical Papers.* Cayley had earlier (Cayley 1843) given a somewhat unoriginal rendition of Rodrigues's paper, as Hawkins (Hawkins 1977) has observed, and in (Cayley 1848) he gave a quaternionic formulation of

[4]Euler was interested in arrays of coefficients $X = Ax + By + Cz$, $Y = Dx + Ey + Fz$, $Z = Gx + Hy + Jz$ which satisfied the conditions $A^2 + B^2 + C^2 = 1$, etc., and $AD + BE + CF = 0$, etc. He presented the array sometimes within the context of simultaneous linear equations where the quantities $X^2 + Y^2 + Z^2$ and $x^2 + y^2 + z^2$ are to be equal, sometimes shorn of the variables as *novem numeros ita in quadratum disponendos* (Euler 1770 p. 288). Similarly for the 4-by-4 case (Euler 1770 p. 313).

[5]The occurrence of half-angles gives a twofold covering of the rotation group; see (Chevalley 1946).

[6]It seems that Gauss invented his quaternions to simplify the composition of rotations in $\mathbb{R}^3$, but he never published his discovery. They are Hamilton's complex conjugates.

other results by Rodrigues. In these two papers he considered their implications for mechanics, thus directing attention back to Euler's concerns. It is striking that he did not feel inclined to appreciate the novelty of the abstract objects here being introduced into mathematics; neither here nor later, when he was studying groups and matrices, did he seek to recall Rodrigues's work. It is perhaps for this reason that Rodrigues is only remembered, if at all, for his formula for multiplying quadruples (θ, g, h, l). When, for instance, Jordan (Jordan 1867, 1868) gave a group-theoretic interpretation of Bravais's (Bravais 1849) work on crystal lattices in terms of discrete subgroups of the group of rigid body motions in Euclidean space, Rodrigues's name is not mentioned. Jordan merely remarked 'One knows that every displacement of a solid body in space is a helicoidal movement and given two such movements one easily constructs the resulting movement which will also be helicoidal[7]

Bravais did not consider motions in space and did not mention Rodrigues. It seems that Klein's confused and possibly second hand description of Rodrigues's work is the only major nineteenth-century source to connect the paper (Rodrigues 1840) with the early history of transformation groups.[8] In any event, Rodrigues's paper is not mentioned in Wussing's thorough survey of the topic (Wussing 1969). Coolidge's earliest reference (Coolidge 1940) to motions as an object of study is (Poinsot 1851). This long paper is, however, chiefly devoted to giving a vivid description of how a solid body moves in space in terms of its concurrent rotations or screw motions about three axes in elliptic cones (Poinsot 1851, Part 3, §37). Most likely Jordan had read this paper. In keeping with the dynamical aims of the paper, forces are resolved into couples producing the screw motions, and Poinsot referred to Euler but not to Rodrigues even though his paper was also published in Liouville's *Journal*.

In other geometries studied at this time only an implicit reference to transformations is found; figures are studied which are unaltered by all projective transformations in (Poncelet 1822) and (Möbius 1827). No reference to combining transformations was made by Chasles in (Chasles 1837*a*), and no work on non-Euclidean geometry discussed the combination of motions for at least another generation. This is in keeping with the preference for invariants over symmetries that marks much of algebra and geometry before about 1870.

Rodrigues has himself become a somewhat shadowy figure, rating no entry in Poggendorff's bibliographies, (Sarton 1957), (May 1973), or the *Dictionary of Scientific Biography*. Kline (Kline 1972, pp. 531, 883) records that he was born in 1794, is the author of Rodrigues's formula for the Legendre polynomials and of various research into differential geometry, notably curvature of surfaces (see also (Reich 1973)), and that he died in 1851. The *Royal Society Catalogue* (1871) lists several papers by him in the *Correspondance de l'École Polytechnique* (1814–1816) and Liouville's *Journal* (1838–1843) but only one on the theme of groups of motions. Bourbaki comments dryly but accurately in a footnote (Bourbaki 1969, p. 164f2)

[7]'It is known that any displacement of a solid body in space is a helicoidal motion and, two such motions being given, one can easily construct the resultant motion which will itself be helicoidal' (Jordan, 1868, p. 229).

[8]I do not know if Klein had read Rodrigues's paper or merely took his information from Cayley. The latter hypothesis is supported by the fact that Klein always associated Rodrigues with combining quadruples and compared his method with those involving quaternions, e.g., (Klein 1926, Vol. 1, p. 186), but he never mentioned Rodrigues in connection with groups of motions.

that Rodrigues discovered his formula shortly before falling into the oblivion of the nineteenth century.

The significance of the paper

Rodrigues's paper certainly deserves, because its thoroughness, to be considered as a paper in transformation groups, specifically on the group of isometries of Euclidean three-space. It is not, primarily, concerned with dynamics or statics, with the causes and manner of motion of rigid bodies. The laws of motion are studied abstractly, with a view to establishing the rules for combining motions. This point of view is precisely that taken later by Klein in the *Erlanger Program* (Klein 1872). There Klein defined a transformation group as a set of motions closed under the operation of taking products; only later did he add the condition that inverses must exist, even then taking the existence of an identity element for granted. Rodrigues gave a complete description of how to combine the two basic motions, rotations and translations. In modern terms, he exhibited the group of proper Euclidean isometries of three-space as a semidirect product of the proper orthogonal group with the group of translations. He proved that the group of translations is commutative, and he treated its elements as vectors, resolving them along the coordinate axes. On the other hand, he was clearly concerned to present the orthogonal group, as noncommutative. Three points are particularly noteworthy: the stress he places on the noncommutativity of this group; the contrast with the infinitesimal rotations, which do commute; and the realization that the infinitesimal motions generate the finite ones.

Rodrigues's concern that the order of composition matters is patent. It is not a mere cautionary aside nor an almost unremarked consequence of the formulae, but it is a matter for him of the greatest significance, reflecting his abstract approach to the subject matter. He was explicitly drawing attention to a novel property of the abstract objects (motions) that he was introducing into mathematics. He stressed in several places that the order of composition matters, before giving the matrix formulae for combining rotations, where it might have been allowed to pass unheralded. The attention he paid to the rules of combination is therefore most interesting.

There is a sense in which it had been well known for many years that matrices do not commute: the programme of simplifying systems of linear equations by change of basis does not make sense, for instance, if all matrices commute. But this awareness was covert, so to speak, and is not the same as an explicit consideration of the rules for combining matrices. It seems that the first person to draw attention to the noncommutative nature of matrix multiplication was Eisenstein in 1844 (quoted in (Hawkins 1977, p. 85)). As Hawkins (Hawkins 1977) has shown, Eisenstein also had the idea of regarding a matrix as a single object and supplied single letter notation for it, which Rodrigues did not, but by preferring to reason geometrically rather than analytically, Rodrigues clearly conceived of each isometry as a single thing, an array.

It is interesting to note that other noncommutative systems were introduced successfully into mathematics at about this time, although earlier attempts had met with less success. Galois' profound remarks on permutation groups (Galois 1832) elicited no response at all for many years, as is well known. It is hard to determine the impact of Abel's remarks, e.g., (Abel 1829), on the connection between the solvability of certain polynomial equations and the commutativity of the group of permutations of their roots, but it would seem that Cauchy's study (Cauchy 1845) is the first time Cauchy, at least, took the conspicuously noncommutative behaviour of permutations seriously; his earlier studies, done around 1815, contain no explicit reference to it. There was a small French tradition of studying symbolic methods in analysis, and it was there that the word commutative was first introduced into mathematics in 1815. Servois said of two functions F and f that they were 'commutatives entre elles' if $F(f(x)) = f(F(x))$ (Servois 1815), but this formal analysis was disapproved of in France for its lack of rigour, as Koppelmann has discussed (Koppelmann 1971). The tradition did much better when transplanted to British soil, and in 1837 Murphy began the study of noncommutative operators in the context of differential equations. This algebraic approach to analysis was greatly refined by Boole (Boole 1844), and its general acceptance into mathematics dates from then.

Grassmann's largely unread *Ausdehnungslehre* appeared in 1844, but undoubtedly the paper most immediately related to (Rodrigues 1840) is Hamilton's 1843 announcement of the discovery of quaternions.[9] As has been remarked, Hamilton came to his discovery algebraically and only put the geometrical considerations first in his later writings. Hamilton's quaternions are numberlike, they are generalizations of complex numbers, and their noncommutativity resounds throughout the subsequent study of algebra. Rodrigues's motions are geometric, but both point toward new, abstract, totalities worthy of the mathematicians' attention. (Further discussion of the implications of Rodrigues's paper on the theory of quaternions may be found in Chapters 2 and 9.)

Indeed, the sudden emergence of noncommutative systems of abstract entities into mathematics in the early 1840s is of considerable historical interest. At the risk of banality, it should be pointed out that all these systems arose from real problems in mathematics. Matrices, at least as arrays of coefficients, abounded in contemporary problems. Permutations derive from the study of polynomial equations, with the unsolvability of the quintic as a central topic. Quaternions are a natural extension of the then best description of real and complex numbers. In each case, except the last, some attention is switched from the study of, or search for, a particular object of a given kind to the study of all objects of that kind; Hamilton's approach was necessarily directed towards the entire algebra. So one could look for the origins of the abstractness of much of modern mathematics in the work of this period. It could also be said that this period saw the emergence of the mathematical profession as a sizable body of people, with substantial schools developing in Britain (Boole was 25 in 1840; Cayley, 19; Hamilton, 35; Sylvester, 26), Germany (preeminently Dirichlet, then 35; Eisenstein, 17; Jacobi, 36; Kummer, 30; and Weierstrass, 25), and Russia, as well as a powerful new generation in France (Hermite, 18; Liouville, 31) to name only a few. It might be that the growth of the profession itself spurred mathematicians to seek abstractions from within mathematics as well as from neighbouring disciplines. Klein in his *Entwicklung* (Klein 1926) suggested that whereas the great mathematical physicists of 1800 belong equally to mathematics and physics, by 1850 a separation of the subjects

[9]There is no reason to suppose Hamilton had read Rodrigues's paper.

is beginning with the appearance of physicists (Maxwell, Helmholtz) whom mathematicians have not been able to claim as their own. Perhaps the gulf was then beginning to inhibit the easy movement between disciplines.

Whatever the validity of these highly tentative speculations, they will not be pursued further here. Rodrigues's paper itself did not draw its audience towards abstraction. Jacobi and Klein saw in it a pleasing calculus, and Cayley redirected it towards problems in mechanics; between them they returned its topic to various aspects of the extensive work of Euler from which the nineteenth century grew so energetically. In so doing they missed its most important idea, which was only to be rediscovered a generation later by Jordan, Klein, and Lie: that of the transformation group.

Bibliography

Abel, N. H. (1829). Mémoire sur une classe particulière d'équations resolubles algébriquement. *Journal für die reine und angewandte Mathematik*, **4**. Also *Oeuvres complètes* (L. Sylow and S. Lie, eds.), **1**, 478-507, Christiania, 1881.

Boole, G. (1844). On a general method in analysis. *Philosophical Transactions of the Royal Society*, London, **134**, 235.

Bourbaki, N. (1969). *Élements d'Histoire des Mathématiques.* Hermann, Paris.

Bravais, A. (1849). Mémoire sur les polyédres de forme symétrique. *Journal für die reine und angewandte Mathematik*, **14**, 141–180.

Cauchy, A.-L. (1845). Sur le nombre des valeurs égales ou inégales que peut acquerir une fonction.... *Comptes Rendus*, Paris, **21**, 779. Also *Oeuvres* (Ser. 1), no. 303, 323–341.

Cayley, A. (1843). On the motion of rotation of a solid body. *Cambridge Mathematical Journal*, **3**, 224–232. Also *Collected Mathematical Papers*, **1**, (1897), no. 6, 28–35.

Cayley, A. (1845). On certain results relating to quaternions. *Philosophical Magazine*, **26**, 141–145. Also *Collected Mathematical Papers*, **1**, no. 20, 123–126.

Cayley, A. (1848). On the application of quaternions to the theory of rotation. *Philosophical Magazine*, **33**, 196–200. Also *Collected Mathematical Papers*, **1**, no. 68, 405–409.

Chasles, M. (1830). Note sur les propriétés generates du systéme de deux corps semblables entre'eux. *Bulletin des Sciences Mathématiques* (1er. section du *Bulletin Universel*), **14**, 321–326.

Chasles, M. (1837*a*). Aperçu historique sur l'origine et le développement des méthodes en Géométrie. *Mém. couronnés par l'Académie Royale des Sciences et Belles Lettres*, Bruxelles.

Chasles, M. (1837*b*). *Mémoire de Géométrie*, published as sequel to (1837*a*).

Chevalley, C, (1946). *Theory of Lie Groups.* Princeton University Press.

Coolidge, J. L. (1940). *A History of Geometrical Methods.* Oxford University Press. Reprinted by Dover in 1963.

Eisenstein, G. (1844). Allgemeine Untersuchungen über die Formen dritten Grades mit drei Variabeln.... *Journal für die reine und angewandte Mathematik*, **28**, 289–374.

Euler, L. (1758). Du mouvement de rotation des corps solides autour d'un axe variable. *Mémoires de l'Académie Scientifique*, Berlin, **14**. Also *Opera Omnia* (Ser. 2), **8**, 154–193.

Euler, L. (1765). *Theoria motus corporum solidorum seu rigidorum—Statica.* Also *Opera Omnia* (Ser. 2), **4**, §690.

Euler, L. (1770). Problema algebraicum ob affectiones prorsus singulares memorabile. *Nov. Comm. Acad. Sci. Petrop.*, **15**, 75–106. Also *Opera Omnia* (Ser. 1), **6**, 286–315.

Euler, L. (1775*a*). Formulae generates pro translations quaecunque corporum rigidorum. *Nov. Comm. Acad. Sci. Petrop.*, **20**, 189–217. Also *Opera Omnia* (Ser. 2), **9**, 84–98.

Euler, L. (1775*b*). Nova methodus motum corporum rigidorum determinandi. *Nov. Comm. Acad. Petrop.*, **20**, 208–238. Also *Opera Omnia* (Ser. 2), **9**, 99–125.

Galois, E. (1832). Lettre à Auguste Chevalier. *Revue Encyclopédique*, Sept. 568. Also *Oeuvres mathématiques*, Paris, (1897), 25–32.

Gauss, C. F. (1819). Mutation des Raumes. *Werke*, **8**, 357–362 (published in 1900).

Grassmann, H. (1844). *Die lineale Ausdehnungslehre.* Reprinted by Chelsea in 1969. English translation: *Linear Extension Theory*, in *A New Branch of Mathematics*, Lloyd C. Kannenberg, Open Court, Chicago and La Salle, 1995.

Hamilton, W. R. (1843*a*). Quaternions. *Mathematical Papers*, **3**, 101–105, Cambridge University Press, 1967.

Hamilton, W. R. (1843*b*). Letter to Graves on quaternions. *Philosophical Magazine*, **25**, (1844), 489–495. Also *Mathematical Papers*, **3**, 106–110, Cambridge University Press, 1967.

Hawkins, T. (1977). Another look at Cayley and the theory of matrices. *Archives Internationales d'Histoire des Sciences*, **26**, no. 100, 82–112.

Jacobi, C. G. J. (1881). Bemerkungen zu einer Abhandlung Eulers über die orthogonale Substitution. *Mathematische Werke* (H. Kortum, ed.). Reprinted by Chelsea, New York, **3**, 599–609.

Jordan, C. (1867). Sur les groupes de mouvement. *Comptes Rendus*, Paris, **65**, 229–232. Also *Oeuvres*, **4**, (1964), 113–115.

Jordan, C. (1868). Mémoire sur les groupes de mouvement. *Ann. di Mat.*, **11**, 167–215, 322–345. Also *Oeuvres*, **4**, (1964), 231–302.

Klein, F. (1872). Vergleichende Betrachtungen über neuere geometrische Forschungen (Erlanger Programm). Also *Gesammelte Mathematische Abhandlungen*, **1**, 460–497.

Klein, F. (1884). *Vorlesungen über das Ikosaeder....* Teubner, Leipzig.

Klein, F. (1926). *Vorlesungen über die Entwicklung der Mathematik im 19. Jahrhundert.* Reprinted by Chelsea in 1967.

Klein, M. (1972). *Mathematical Thought from Ancient to Modern Times.* Oxford University Press.

Koppelman, E. (1971). The Calculus of Operations and the Rise of Abstract Algebra. *Archive for History of Exact Sciences*, **8**, no. 3, 155–242.

Lagrange, J.-L. (1788). *Mécanique analytique* (3rd ed., 1853) (J. Bertrand, ed.).

May, K. O. (1973). *Bibliography and Research Manual of the History of Mathematics.* University of Toronto Press.

Möbius, A. F. (1827). *Der Barycentrische calcul....* Leipzig.

Monge, G. (1784). *Mémoires de l'Académie Royale*, Turin.

Poinsot, L. (1851). Théorie nouvelle de la rotation des corps. *Journal de Mathématiques*, **16**, 9–129, 289–336.

Poncelet, J. V. (1822). *Traité des propriétés projectives des figures.* Paris.

Reich, K. (1973). Die Geschichte der Differentialgeometrie von Gauss bis Riemann. *Archive for History of Exact Sciences*, **11**, 273–382.

Rodrigues, O. (1840). Des lois géométriques qui regissent les déplacements d'un systéme solide dans 1'espace,.... *Journal de Mathématiques Pures et Apliquées*, **5**, 380–440. *Royal Society Catalogue of Scientific Papers, 1800-1863*, **5**, (1871), 250–251.

Sarton, G. (1936). *The study of the history of mathematics and the study of the history of science.* Harvard University Press. Reprinted by Dover in 1957.

Servois, F.-J. (1815). Essai sur un nouveau mode d'exposition des principes du calcul différential. *Annales des Mathématiques*, **5**, 98.

Wussing, H. (1969). *Die Genesis des abstrakten Gruppenbegriffes.* Berlin. English translation: *The Genesis of the Abstract Group Concept.* Translated by Abe Shenitzer with the editorial assistance of Hardy Grant, MIT Press, Cambridge, MA, 1984.

CHAPTER 9

After Rodrigues: From Rotations to Quaternions

EDUARDO L. ORTIZ

Introduction

As in most countries, the reception in France of Hamilton's quaternions throughout the second half of the nineteenth century totally obscured the significance, in this respect, of Rodrigues's work on rotations. In the first part of this article we consider the impact of the quest for geometrical representations of complex numbers on attempts to formulate an abstract language for operations on geometrical figures, in some ways similar to that of algebra. The realization of the limitations of the language proposed by the Italian mathematician Bellavitis to translate from problems in the plane into those in space led to the consideration of quaternions as a more suitable concept in the machinery to implement that language. However, with the acceptance of quaternions also came the acceptance of the loss of commutativity and, with the latter, the loss of a vital contact with the language of ordinary algebra.

It is interesting that the main character in the first phase of this story is a friend of Olinde Rodrigues, Abel Transon, a French mathematician, a polytechnicien, and a Saint-Simonian, who believed firmly in the need to transform society. Although a respectful friend of Rodrigues, he did not realize the implication of Rodrigues's work on the theory of quaternions, which shows that the connection between these two methodologies was overlooked even within Rodrigues's closest circle of friends.

The second half of the nineteenth century was disrupted in France in 1870 by her defeat at the hands of Prussia, which began a period of introspection and of reconstruction for many of its institutions. Concerns were not limited to society: some scientists, including mathematician Laisant and his close friends Lemoine, Laurent, Barisien, Brocard, Laussedat, Cornú, Koenigs, shared the belief that some basic changes were necessary in the structure of the French scientific community in order to bring about in their country an *avancement* of science almost on an industrial scale, which they thought was necessary.

They believed that the organization of science in France had an antiquated structure and that the channels of communication with other scientific communities in Europe, particularly with the German-speaking world, were too narrow. Some of them also believed that the interaction between science and society was too weak, if compared with what, in their perception, was the situation in Germany. They went as far as listing a lack of coupling between science and the army in France as one of the ingredients of the Prussian victory.[1]

The institutional developments they sponsored nurtured closer scientific relations between Paris and the rest of the country, as much as between French institutions and foreign scientific societies. Later, Laisant and his friends embarked on new projects to promote wider channels of global scientific communication, focused on the consolidation of an international mathematical community. In sharp contrast with earlier periods, new societies were created in rapid succession for the promotion of science in France, either in general or in specific branches or in the interaction between science and its social context. The activities of the group of Laisant's friends were responsible for the creation of the *Société de Physique*, the *Société Mathématique de France*, and the *Assoçiation Francaise pour l'Avancement des Sciences*,[2] today still pillars of the French scientific community. Their sense of planned development is an idea that draws as much from the Revolutionary period as it does from the ambitious projects for the advancement of society of Saint-Simon and Charles Fourier.[3]

The early contributions of Olinde Rodrigues do not seem to have constituted a distinctive element in the second accelerated phase of the process of the acquisition of Hamilton's quaternion theory in France. Crossing the Channel to France was a slow process for quaternions, which shows some remarkable fractures and historical discontinuities. Laisant, and other members of his group, played an important role in this last stage of transmission. In addition, quaternions were for them a further example of the French mathematical community's reluctance to look outside its own frontiers.

Only towards the turn of the century did there resurface a wider understanding of the fact that the early work of Rodrigues on rotations in space around a fixed point was closely related to Hamilton's concept of quaternions.

The algebraization of geometry: Accomplishments and failures

The work of the Italian mathematician Giusto Bellavitis ((Freguglia 1991) and (Ortiz 1999)) had some influence in awakening the interest of French mathematicians to the vast possibilities opened up by ideas associated with vectors and, later, with quaternions. Bellavitis was a representative of the classical geometrical tradition (Knight 1968), which by the middle of the nineteenth century had persisted more firmly in Italy than in the upper echelons of the French mathematical community. Curiously, Bellavitis's route did not cross the path of an even earlier French tradition on topics related to the concept of quaternions, the composition of angular rotations about a point in space, of which Olinde Rodrigues was an outstanding contributor, as Gray (Gray 1980) and Altmann (Altmann 1989) have shown.

In a series of important papers going back at least to 1835,[4] Bellavitis reconsidered the ideas of Carnot (Carnot 1803) and the earlier work of Möbius (Möbius

[1]In this respect the lecture on war and science, given by the polytechnicien Colonel Aimé de Laussedat at the first meeting of the *Assoçiation Francaise pour l'Avancement des Sciences* is particularly significant. See (Ortiz 1996, pp. 332-333).

[2]On the history of the last two societies, see the well-documented work of Hélène Gispert (Gispert 1991, 2002).

[3]These conceptions affected a number of other disciplines, included urban design; see (Benevolo 1963).

[4](Bellavitis 1835, 1838, 1854, 1858).

1856, p. 227) on imaginary forms in algebra, which he applied when discussing the possibility of introducing into geometry an algorithm capable of dealing, equally, with magnitude and position of different parts of a figure.[5] Ideally, the algorithm should operate following the patterns established by ordinary algebra, for some time regarded as the most sophisticated form of a scientific analytic language (Ortiz 2003). Applications of this calculus could be extensive and have an impact not only on geometry but also on graphical calculations and mechanics, both of which were, respectively, branches of a considerable practical and theoretical importance in the period.

Bellavitis placed an equivalence relation, the notion of *equipollence*, at the basis of his symbolism. This concept enabled his calculus to deal with classes of geometrical objects rather than with individual objects. Geometrical objects were regarded as equivalent provided that one was the result of a fixed parallel displacement of the other. The notion of equipollence was intended, in fact, to put parallel displacements into a geometrically subordinate position. Breaking the dependence on specific systems of coordinates or on arbitrarily chosen elements, Bellavitis's calculus, into which parametrization could be easily introduced, brought considerable simplification to the treatment of the theory of curves and surfaces.

The geometric interpretation of complex numbers is a topic that attracted a great deal of attention in the first third of the nineteenth century. One of the interesting characteristics of the symbolic language that Bellavitis proposed was his readiness to give complex numbers a central role in the construction of his algorithms, realizing that real numbers may not have a sufficiently flexible structure to serve as a mechanism for the language he was trying to devise.

In some ways, he inverted the problem of geometric interpretation by using complex numbers as a tool in the design of a calculus applicable to equipollent figures.[6] On a deeper level, his was an attempt to find a language better suited to the solution of geometrical problems in which the role of geometric intuition would be considerably less central than it is in classical geometry. In this respect it touched on the controversy between analytic and synthetic approaches: reasoning involving chains of appropriate choices of geometrical theorems could be greatly simplified and set in an orderly sequence. With this language, the solution of geometrical problems would become more systematic and methodologically more in line with algebraic reasoning. The internal logic of the language would suggest the next step, without having to return repeatedly to the body of known geometrical results in search for an answer. For this language to be practical, clearly, it had to be structured on the basis of a limited number of rules.

It was soon realized, however, that his interesting method had a very fundamental limitation: it could only be applied to the solution of problems of plane geometry. Furthermore, it did not suggest a clear way out for an extension to three-dimensional space: numbers more sophisticated than complex numbers were required to deal with problems in space.

After intensive efforts over a considerable period of time, Bellavitis realized there were essential constraints in his own language-method and began looking for new possibilities, outside the realm of complex numbers. In 1858 he published a long paper on the connections between his calculus of equipollences and Hamilton's

[5]Bellavitis refers to them as *grandeza e posizione*.

[6]Laisant has made this observation in (Laisant 1881, p. 346).

theory of quaternions (Bellavitis 1858). He clearly perceived that the latter was a far more formidable instrument than his equipollences[7], and by then his ideas had made a serious impact in France.

Roughly at the same time that Bellavitis began to move out of equipollences, his French followers accompanied him into the new territory of quaternions that he was beginning to explore. Though French mathematicians made a point of signaling the presence of French precursors to Bellavitis's ideas, in particular in reference to the geometric interpretation of complex numbers, they did not revisit the ideas of their compatriot Rodrigues.

The early reception of equipollences in France

Bellavitis's ideas were discussed in France by Abel Etienne-Louis Transon (1805–1876)[8] in the second part of (Transon 1868a), a paper on the separation of roots of algebraic equations published in *Nouvelles Annales de Mathématiques.* Transon, however, did not refer to Bellavitis's work in his paper, which he then ignored. Five years later, in a note to a paper published again in the same journal,[9] which was then one of the most interesting French sources of mathematical innovations, Transon indicated that he had been informed that Bellavitis had done similar work independently in Italy. He gave Bellavitis full credit for his work, while emphasizing earlier French contributions to the geometric representation of the 'so unjustifiably called *imaginary numbers.*'

Transon did not approach complex numbers as a pair of real numbers, $a+ib$, but as directed numbers,[10] defined by an angle and a modular distance to the origin of a system of polar coordinates. He believed that directed numbers had been implicitly introduced by Cauchy in a paper read at the *Académie des Sciences.* In turn, according to Transon, Cauchy's ideas were based on the earlier work of a group of French mathematicians: l'Abbé Buée, Jean Robert Argand, C.-V. Mourey, Ambroise Maurice Claude Christophe Faure, and François Vallès. In his paper Transon gave his subject of research the name of *Calcul directif* and, later, *Algèbre directive.* He applied his new symbolism to the study of complex roots of algebraic equations $f(z) = 0$ defined on the plane. Whereas Descartes had constructed analytic geometry using positive and negative numbers, Transon believed that his *Calcul directif,* which gave complex numbers a new role, was a much more powerful tool (Transon 1868*b*, p. 146).

Transon stated that through this mathematical work he had attempted to discuss, on a concrete situation, a much larger metaphysical question: the justification of the idea that falseness cannot contain truth. This statement is based on a concern with the fact that when operations on complex numbers lead to real numbers, they invariably give correct results; thus, they could not be characterized, as in current mathematics textbooks, as 'symbols deprived of any significance.'[11]

---

[7]A few years before his death, in an extensive treatise, Bellavitis (Bellavitis 1876) discussed the origins of his method of equipollences.

[8]The date of birth given in the reliable work of May (May 1973) is incorrect.

[9](Transon 1873). See note on p. 8.

[10]Also called *longueurs inclinées*; (Transon 1868*a*, in note to p. 57).

[11](Transon 1868*b*, p. 146). He probably alluded to Prouhet's remarks in his edition of Lacroix' *Éléments d'Algébre,* but similar words had also been used by Condillac in *La Lange des Calculs,* first published in (Arnoux and Mousier 1798).

In 1869, that is, immediately after the publication of the first of Transon's two papers and in the same journal, Jules Hoüel (Hoüel 1869*a*) published a review of Bellavitis's theory of equipollences and made it widely known in France. In his important and generous review, Hoüel made reference to Hamilton's quaternions and averred that a special algebraic algorithm was required to deal with them. Hoüel's work clearly enjoyed some recognition as, in the same year, it was also published in Paris in the form of a short book (Hoüel 1869*b*).

Transon and Olinde Rodrigues's circle of friends

Besides working in areas of research connected with a discussion on the nature, geometric interpretation, and novel applications of complex numbers, as well as the relation between analytical and synthetical methods, Transon was also concerned with matters bordering philosophy, metaphysics, and politics. As was his respected friend Olinde Rodrigues, Transon was a relevant personality in the history of the Saint-Simonian movement.

Transon entered the *École Polytechnique* in 1823 (Fourcy 1828, p. 468), the year when Rodrigues met Saint-Simon. After graduating in 1825 at the top of the school's rank, he joined the selected world of the *École des Mines*[12] but did not take a state position in mining, as he may have been entitled to do. Instead, he wished to contribute to the creation of a new social system, more firmly based on justice and in which science and industry would contribute to improve the human condition. Transon became a follower of Saint-Simon and later, for a limited period, of Enfantin.

He was not alone in the decision to take this route; others in the school, 'the talented élite' (Tournaire 1878), were also among the young polytechniciens who followed the doctrines of Saint-Simon. Among them were Jean Reynaud,[13] Henri Furnel, and Chevalier[14]. The acceptance of the new ideas was based on the premise that the foundations of ancient society were irreparably broken down and that a new order had to be established. The training in science and industry that the polytechniciens had received made them receptive to the idea that science could help provide the foundations for a new and better society, as well as being an irreplaceable tool in the fulfillment of this purpose.

Transon became a member of the editorial board of the influential Saint-Simonian periodical *Le Globe*, which, before the Paris Academy of Sciences started publishing its own *Compte Rendus*, did not neglect its duty to inform its readers of the latest scientific news[15]. Transon seems to have become one of the most brilliant lecturers within the Saint-Simonian group of polytechniciens. Among his most remarkable public lectures, five addressed to the students of the *École Polytechnique* are specially remembered (Transon 1830).

---

[12]Gabriel Lamé (in 1817), Transon (in 1825), Joseph Bertrand (in 1841), and later Henri Poincaré are among the few *corpsard-Mines* admitted "to those classed for the services of mines" *hors concours*. I with to thank M. Udai Venedem for this reference.

[13]Jean Ernest Reynaud (1806–1863).

[14]Michael Chevalier (1806–1879), who later became a leading economist; his brother August helped collect Galois' works, after his death in 1832.

[15]See (Zerner 1991, pp. 308–310).

Transon's break with Enfantin

In 1832, after a split in the Saint-Simonian movement, Transon left it in the company of Pierre Leroux (1797-1871)[16] and other eminent members of the new religion, including Olinde Rodrigues. In this connection he wrote a *Simple Écrit* (Transon 1832), an important document that seems to have escaped the attention of Saint-Simonian scholars. In it he explained the reason for his leaving *Père* Enfantin, whom he described with sad affection as 'a man that has always treated me as a son,' and made an account of his contributions to the movement over the last two years (Transon 1832, p. 2).

His document reflects his deep disappointment with a movement that he had expected to formulate a new definite religion, already announced by Saint-Simon in his *Nouveau Christianisme* of 1825. Thus he proposed a new concept of industry and contributed to formulate a 'general science' to serve as a 'link to all sciences,' returning to the *Idéologues'* dream of unitary science (Ortiz 2003).

Transon declared that he was tired of waiting for something better than announcements. However, it was when considering the lack of actual results that he confessed that 'my illusions have gradually evaporated.' The departure of Jules Levachelier and his acquaintance with the works of Charles Fourier contributed to deepen his disenchantment with Saint-Simon, whose intellectual achievements he saw in a new light.[17] Particularly, he distrusted the master's thoughts on procedures of association, the peaceful organization of work, and the question of intimate relations between women and men (Transon 1832, p. 3). He also accused Enfantin of practical failures. The workshops for tailors, seamstresses, and cobblers, the projects of common housing for workers in different districts of Paris, the recruitment of a 'pacific army' of workers of all kinds, including artists and engineers, cooperating in large industrial enterprises in which different professions would communicate and interact positively in new ways, had all ended in failure. By then he perceived Saint-Simonians as a dogmatic sect; he opposed Enfantin's ideas in favor of Fourier's better thought-out perspectives.

He also quoted the wise and practical statements of Olinde Rodrigues in connection with their relations with banking and finance: 'he knows, by experience, that one can approach bankers beneficially only on *money matters*, and that these men would not become closer to the doctrine unless there is *money* to be made.'[18] He also accused Enfantin of having forgotten that Christian morality had established rules relevant in the field of industrial activity and that the liberation of women and the organization of industry were closely interwoven matters.

On a more personal line of attack, he accused Enfantin of having gotten rid of Saint-Amand Bazard and later Olinde Rodrigues to clear his way in order to sustain, against Christian and philosophical morals, his own personal views: that everything man does is a valid action. He also quotes respectfully from the works of Olinde's brother, the young Eugène Rodrigues. Eventually, Transon returned

[16] Leroux introduced the word socialism in France, as opposed to individualism; he also founded *Le Globe*, with Paul François Dubois. Later he contributed to the foundation of the *Revue encyclopédique* and of the *Encyclopédie nouvelle*, the latter in the company of Jean Reynaud.

[17] As expressed in his last meeting with Enfantin.

[18] (Transon 1832, p. 5). In this connection, Transon made reference to notes in *Le Globe*,30 December, that year.

to Catholicism and to mathematics and became a senior member of the *École Polytechnique.*

The evolution of quaternion studies in France

We shall consider in the rest of this article a series of mathematicians whose studies gradually brought the subject of quaternions into France, as well as some of the conceptual problems they faced in the introduction of quaternions in mathematics. All these studies consistently ignore Rodrigues's work for, as we have already averred, it was only at the turn of the century that Olinde's work was finally recognized in his own country and then linked to the developments made abroad at a later date.

The younger mathematician Guillaume Jules Hoüel (1823-1886), born in Thaon, Calvados, on April 7, 1823 (Halsted 1897), graduated from the *École Normale* and took his doctorate at the Sorbonne in 1855. In 1859, after a series of different assignments, he was given a chair of pure mathematics at the *Faculté des Sciences* of the University of Bordeaux and remained in that city for the rest of his life. He also played a relevant role in the diffusion of new geometric ideas and an even more important one in the study of the theory of complex variables in France. Hoüel became acquainted not only with the work of Bellavitis but also with that of Hamilton, even in areas outside quaternions: he used Hamilton's results in the analysis of the perturbations of Jupiter, a topic he discussed in his second doctoral dissertation (Hoüel 1855). He was also the author of a widely used textbook on infinitesimal calculus, published in four volumes between 1878 and 1881. Hoüel was likewise remembered for his remarkable collection of tables of logarithms (Hoüel 1858), which were the standard choice for users for nearly a century.

Hoüel, as Transon, had a philosophically oriented mind. However, he differed from Transon not only in the school in which he received his training but also in his religious outlook. Hoüel came from an old Protestant family of the northern part of France. Proficient in several European languages, he translated into French the papers of several leading contemporaneous German mathematicians, notably Dirichlet, Kummer, Krönecker, and Riemann. He also wrote on the principles of geometry and played a remarkable role in making the works of Lobatchevsky and Bolyai available in the French language and through it into contemporaneous intellectual Europe. Contrary to Hoüel, Transon rejected the new theory of parallels.

Quaternions reach the University of Paris: Allégret's doctoral thesis

Hamilton's work on quaternions does not seem to have been discussed in great detail in France until Alexandre Allégret published his doctoral thesis in Paris in 1862 (Allégret 1862). His first thesis was an essay on quaternions; the second was a study on the irregularities in the movement of Jupiter's satellites. The first thesis consisted of three parts. In the first he discussed the rules of the calculus of quaternions, in the second he gave a geometrical interpretation of the new mathematical objects, and in the third he gave applications to the theory of lines and surfaces. In the last quarter of his work (Allégret 1862, p. 45), as an application of the calculus of quaternions, he derived the main results of spherical trigonometry, a subject closely connected with position astronomy and also a good opportunity to revisit Rodrigues's work, which he, however, does not seem to have known about.

Although approved by a *Commission d'Examen* presided over by Chasles and composed of Puiseux[19] and Serret,[20] Allégret's work was not well received by some established mathematicians. Eugène Prouhet, well known for his successful efforts to revive Lacroix' series of mathematical textbooks,[21] became editor of *Nouvelles Annales de Mathématiques* in 1863 and remained in this position until his death in 1867. In his first year as editor of that journal, he inserted a note with his views on Allégret's thesis (Prouhet 1863). His review was critical, as he perceived quaternions as wholly artificial mathematical objects. Prouhet remarked: 'To invent expressions that in themselves offer no meaning to the spirit, and to subsequently search one such by means of what one calls a geometric interpretation, is no different from, after having constructed a beautiful sentence, one were to look for a thought that could be assigned to it.' Taking into account the date of these remarks, it is quite possible that Prouhet had in mind a paper that Hamilton had read at the Dublin meeting of the *British Association for the Advancement of Science* a few years earlier, in which he introduced his new icosian calculus (Hamilton 1856). In this paper, Hamilton introduced new roots of unity, justifying this invention by giving, precisely, a useful geometrical interpretation of them[22].

Prouhet stressed that the method was chiefly recommended by Hamilton's prestigious name and commented that Allégret's essay had as much clarity as 'the subject can have.' Prouhet also commented on Allégret's derivation of the spherical trigonometry theorems, complaining that he was not impressed by the use of a rather complex machinery to prove only elementary results. As with Allégret, Prouhet makes no reference to Rodrigues either.

The diffusion of Bellavitis's work in France: Charles-Ange Laisant

Hoüel's review paper on equipollences had a considerable impact. A young French mathematician, Charles-Ange Laisant, decided to follow it up with a direct translation of Bellavitis's 1854 Modena book on equipollences. His translation, which included extensive notes and comments, was serialized by *Nouvelles Annales de Mathématiques* in 1873–1874 (Bellavitis 1873) and was immediately published in Paris as a book (Bellavitis 1874) by Mallet-Bachelier's successors in the science-publishing business, the new firm of Gauthier-Villars. While working on the translation, Laisant corresponded with Bellavitis, from whom he acknowledged to have received advice[23].

Laisant was born in 1841; he entered the *École Polytechnique* and later the *École d'Application* in Metz. He had a genuine mathematical interest and contributed to a number of mathematics periodicals, geometry being his main subject. He also showed a deep commitment to mathematics teaching and to its advancement, as we shall see later. Laisant became an *Examinateur d'admission* and a *Repétiteur* at the *École Polytechnique*.

[19]Victor Alexandre Puiseux (1820–1883) was the current professor of astronomy.

[20]Joseph Alfred Serret (1819–1885) was professor of infinitesimal calculus at the Sorbonne.

[21]Eugène Prouhet (1817–1867), a *Répétiteur* at the *École Polytechnique*, added notes to Lacroix' series of mathematical textbooks and adapted them to successive changes in the syllabus of the entrance examination for the *École Polytechnique*. His revised editions extended the life of Lacroix' books until, at least, the end of the century. See (Nový 1973, pp. 63–68).

[22]For a discussion on the mathematical significance of Hamilton's work on octonions, see the interesting paper of Baez (Baez 2001).

[23]Laisant's notes in (Bellavitis 1874, p. xiv); see also (Laisant 1880, p. 343).

Although he could not be regarded as a leading member of the French mathematical research élite of his time, he was active and well connected. His contemporaries describe him as warm and generous, which many a page of his work as well as some of his actions suggest to be an accurate description. Laisant died in Paris in 1920, deeply saddened by the shattering of internationalist ideas as a result of the First World War.

As indicated before, the geometrical basis of the calculus of equipollences seemed to make it equally relevant in a reciprocal formulation, that is, as a tool to provide analysis and algebra with geometrical elements more closely related to immediate, visual, reality. According to Laisant, Bellavitis's calculus could provide these disciplines with 'real geometrical objects in place of imaginary symbols.' He shared with a group of his mathematical friends the idea that complex numbers ought to be reinterpreted, possibly with the help of geometry, to bring them back to the 'real' world of visual 'realities.' Insistence on bringing a sense of realism into the field of complex quantities did not prevent Laisant, as it did not prevent Hoüel either, from joining with enthusiasm in the exploration of the new non-Euclidean geometries, as a possible, even if not 'real,' mathematical theory.

In between the Italian and French editions of Bellavitis's Modena book, Hamilton's research on quaternions became better known outside the English-speaking world. In a paper mentioned earlier, Bellavitis (Bellavitis 1858) tried to relate his equipollences to Hamilton's quaternions, accepting that quaternions allowed for a three-dimensional extension that had eluded him when using only equipollences. But Bellavitis's program of replacing algebraic and analytical symbols with objects carrying a deeper, geometrical, sense of 'realism' stumbled upon a serious logical difficulty: when insisting on their extension to space, these objects would not carry with them the ordinary rules of algebra. This was a very high price to pay and a difficult step to take.

From Bellavitis to Hamilton: A reevaluation of Mourey's work

Laisant underlined the importance of the early work of Mourey (Mourey 1828) on the theory of inclined lines and of the idea of *nombres directifs* from the viewpoint of geometrical applications. He also remarked on the influence Mourey had on Transon's work, placing Mourey firmly as a precursor to Bellavitis. In his view this earlier work did not diminish Bellavitis's merits as the creator of an algorithm for the application of the principles of algebra to plane geometry[24].

Resistance to accept the new ideas of Bellavitis, and later those of Hamilton, was considerable and not simply confined to France. It also existed in Great Britain with some notable exceptions. In his address to the Mathematical and Physical Sections of the *British Association* in 1870 Maxwell (1890, II, p. 220) communicated to his audience positive views on the possible application of quaternions to physics: 'the Calculus of Quaternions, a branch of mathematics which, when it shall have been thoroughly understood by men of the illustrative type, and clothed by them with physical imagery, will become, perhaps under some new name, a most powerful method of communicating truly scientific knowledge to persons apparently devoid of the calculating spirit.' Later, moreover, for the *London Mathematical Society*, he compared the importance of the new method with the introduction of coordinates by Descartes (Maxwell 1890, II, p. 259). Tait, however, who was to

[24]Laisant's notes in (Bellavitis 1874, p. xiii).

become the champion of quaternions after Hamilton's death, could not convince his coauthor Thomson to introduce quaternions in their famous joint textbook on natural philosophy[25].

In France, Laisant seems to reinforce this view through his effort to use references to French mathematicians, to bring home the fact that the ideas leading to Bellavitis's work and later to the invention of quaternions had, indeed, French roots. Laisant expressed amazement at French mathematicians continuing to ignore a method 'that originated first amongst us.'[26] He also did not look, however, sufficiently close: as Rodrigues's contributions continued to be ignored.

Hamilton (1899, I, p. 276) also referred to Mourey's book, but he did it in a very different tone. For Hamilton, Mourey's was a 'very ingenious and original little book,' from which he quoted a very specific result. However, the question of Mourey's priority may have been debated before Laisant's writings appeared. In connection with quaternions Hamilton was forced to remark that 'Mourey's book contains *no hint* of the *present calculus*, being *confined*, like the *Double Algebra* of Prof. De Morgan (London 1849), and like the earlier work of Mr. Warren (Cambridge 1828), to questions within *the plane*; whereas the very *conception* of the *Quaternions* involves, as we have seen, a reference to *Tridimensional space*.'[27]

Laisant's thesis on quaternions at the University of Paris

The climate of opinions on quaternions was changing in France in the 1870s. After 1870, nearly ten years after the publication of Allégret's thesis, Hoüel regularly reviewed books related to quaternions in the pages of the *Bulletin des Sciences Mathématiques et Astronomiques*[28]. As we have already indicated, the fourth volume of Hoüel's treatise on complex quantities, devoted precisely to quaternions, appeared in 1874. Publication of the treatise had started in 1867, when Hoüel was thirty-four years old. The last part was so successful that it was also published as a separate volume (Hoüel 1874). It was not unexpected, therefore, that when in 1877 Laisant submitted two theses to the University of Paris for his *doctorat ès-sciences*, the first one dealt with applications of quaternion calculus to mechanics [29]. On 29–30 November 1877, his work was examined by a committee presided over by Briot, the leading expert in complex functions, who was then professor of probability and of mathematical physics in Paris; Bonnet[30] and Darboux[31] were his coexaminers.

Darboux's and Briot's views on the thesis[32] and more generally on the theory of quaternions have some interest. In his report written on 30 November, Darboux remarked that quaternions, as a 'subject has not yet been treated in France,' while

---

[25]See (Knott 1911, p. 185) and also (Smith and Norton Wise 1989, pp. 360–372).

[26]Laisant's notes in (Bellavitis 1874, p. xv).

[27]Hamilton refers to (De Morgan 1849) and (Warren 1828).

[28]G. Dillner's 1860 book *Geometrisk kalkyl eller geometriska quantiteters räknelagar* in Vol. **1**, 1870, p. 249; the second edition of Tait's *An elementary treatise on quaternions*, in Vol. **6**, 1874, p. 161, and P. Romer's 1876 book *Principles fondamentaux de la méthode des quaternions*, in Vol. **11**, 1876, p. 113.

[29](Laisant 1877). *Thèse*, No. de Ordre 297. Faculté des Sciences, 133 pp., Paris. *Premier thèse: Applications mécaniques du calcul des quaternions.*

[30]Pierre Ossian Bonnet (1819–1892) was then regarded as one of the senior geometers of France; he received the *Grand Prix des Sciences Mathématiques* in that same year.

[31]Jean Gaston Darboux (1842–1917) was a younger but already well-known geometer; he later became director of studies at the *École Polytechnique*.

[32]*Archives Nationales*, Paris, Séries AJ^{16}/5533 (1870–1880).

Briot indicated that '[t]he method of quaternions, invented by the English [sic] geometer Hamilton, is still little known in France.' Darboux mentioned Allégret's thesis and Hoüel's work and pointed to the applications of quaternions as the main merit of Laisant's thesis. In praising his effort for 'the work that he must have done in order to assimilate such a difficult theory,' the examiner gave a clear indication of his views on the difficulties associated with the theory of quaternions. Still then, Laisant regarded that theory mainly as a spatial extension of Bellavitis's calculus, requiring the introduction of special algebraic rules.

In the section of his thesis dealing with quaternions, Laisant made a reference to their reception in France. He indicated that it was already twenty years since Hamilton had published his book on the theory of quaternions and stated that such theory had attracted the attention of a number of mathematicians in Great Britain, Italy, Germany, and America. He regretted that since 1862, the date of Allégret's thesis, no mathematician had done any substantial work on quaternions in France.

In his controversial introduction, Laisant questioned the path of mathematical research in France and quoted and queried the wisdom of Prouhet's indictment of Allégret's work. He accepted that the spirit of Prouhet's criticism was generally valid, but he remarked that it did not apply to quaternions. Playing with Prouhet's words on Allégret's dissertation, he quipped: 'it is not to make a phrase in order to ask oneself later what it can signify.'

There is a direct link between Laisant and Hoüel's interest in quaternions. In the introduction to his thesis Laisant named Hoüel as his mentor in quaternions. Furthermore, he indicated that in his thesis, for the general theory, he had relied on Hoüel's work. His treatment of the latter was brief, and he soon embarked on his proper topic, mechanical applications. Laisant does not claim full originality for his work but only for the way in which some of his results on the application of quaternions to kinematics, static, and dynamics were derived. However, Darboux, in his report, states that there are 'new insights' which relate to the problem of the rotation of a solid about a fixed point, again with no reference to Rodrigues.

On quaternions, however, even the faithful had second thoughts. Answering in advance to possibly negative comments, Laisant reassured his readers that the theory of quaternions could not supersede other methods already used in rational mechanics, as quaternions were not of universal application. He believed, however, that they offered a number of advantages, including simplicity, uniformity of approach, and the fact that the solution of mechanical problems is somehow simplified by having to consider only one equation. Prouhet (Prouhet 1863, p. 334) had also been prudent when dealing with the future; commenting on Allégret's thesis he warned: 'We do not know what the future keeps for quaternionic analysis.'

Quaternions as an integral part of France's mathematical scenario

If in his dissertation Laisant relied on Hoüel's treatise for the general theory, this work was excessively long and, perhaps, not particularly suited for the beginner in the seven years that had passed. In 1881 Laisant published a basic textbook on quaternions (Laisant 1881); it contains what had been left out of his thesis: an introduction to the general theory. He indicated in the preface that writing this book was the result of the insistence of some of his friends. They believed that the quaternions 'cause' could be enhanced if an easily accessible and concise introductory textbook became available.

Laisant (Laisant 1881, p. v) also claimed that he wrote this book over the three years after the presentation of his doctoral dissertation and that it was written in the few moments left to him by his then current duties. He described them as 'numerous and totally alien to mathematics,' a remark whose significance will later on transpire.

Laisant defined Bellavitis's work as the calculus of imaginary quantities in the plane, to which the ordinary rules of algebra apply, and indicated that several attempts had been made to extend these ideas to space. He defended Hamilton's theory of quaternions, which in his view was 'a new doctrine, unimpeachable from the philosophical point of view, fertile in results' (Laisant 1881, pp. vii–viii). Laisant defined quaternion calculus in Bellavitis's terms, as he had already done in his translation of Bellavitis's work[33]: the algebra of geometrical objects in space.[34] Altmann has shown, however, that this view of quaternions cannot be accepted uncritically.[35]

Laisant's was not the only attempt to make quaternions more widely known in France; roughly at the same time the French Alsatian mathematician Gustave Plarr, who had sought exile in England after the Franco-Prussian War of 1870, was translating into French the second edition of Tait's elementary treatise on quaternions (Tait 1867) for Gauthier-Villars. The first volume, on the theory of quaternions, appeared in 1882, a year after Laisant's book; and the second volume, on applications to mechanics and mathematical physics, appeared two years later (Tait 1882–1884).

The loss of commutativity: Theoretical questions posed by quaternions

As geometrical objects, quaternions admit the ordinary operations of addition and subtraction, but when their multiplication rules were defined, it was found that there was some freedom. If a division algebra is to be retained,[36] it becomes necessary to define for the basic units i, j, k a multiplication table in which $i^2 = j^2 = k^2 = -1$, $ij = k$, while $ji = -k$, and so on by cyclic permutation. If the more symmetric law $ij = ji$ were adopted, it is easy to find nonzero quaternions whose product is equal to zero. However, that multiplication table breaks the trivial analogy with the algebra of complex quantities.

When referring to the loss of commutativity in the new algebra of quaternions, Laisant was expressing a negative feeling shared by many of his contemporaries, even among those who recognized the importance of quaternions. He remarked that the loss of commutativity created 'especial difficulties'[37] as it made the theory of quaternions difficult and, from a critical point of view, introduced a fundamental and serious problem. He stressed, however, quite rightly, that this restriction was in the very nature of the questions discussed and a direct consequence in the calculus of quaternions of the intrinsic geometrical characteristics of three-dimensional space:

[33] Laisant's notes in (Bellavitis 1874, p. xv.)

[34] Laisant's notes in (Bellavitis 1874, p. viii).

[35] See (Altmann, 1989, pp. 300–305), where the problem of pure quaternions as three-dimensional vectors is discussed. This source is closely followed in our discussion on quaternions in the next paragraph. See also Chapter 2.

[36] That is, an algebra in which the product of two elements vanish if and only if one of them is the null element.

[37] Laisant's notes in (Bellavitis 1874, p. ix).

'But this difficulty comes from the very nature of things. It represents the exact transliteration, formally, of a precise fact.'[38]

Laisant, again, made reference to secondary aspects of the theory to make it more acceptable. The wide range and importance of the results of applications should reassure his readers that the new algebra of quaternions that they were offered deserved to be considered and accepted. In addition, there was a notational-symbolic aspect associated with quaternions: they made arguments more concise. He also claimed that they provided a sense of intuition, a very personal, individualistic remark that was difficult for him to define more precisely than that. Laisant warned the students of quaternions against symbols leading them to lose track of realism: he invited them 'not to separate for a single instant the symbols from the real facts of which they are the representation,' and to be anchored to the reality of geometry (and possibly of mechanics) through an awareness of the geometrical meaning of what they are trying to express or combine.[39]

Quaternions and the jealousies between analysis and geometry

In his book, now a full text available to French readers in their own language, Laisant did not hesitate to enter into controversies of a more general character. He tried to elucidate the reasons for the long delay in the acceptance of quaternions by French mathematicians, if compared with their British or American colleagues. He explained them in terms of the exclusively analytical way in which quaternions had been presented to French readers before, although Laisant does not identify these presentations. Clearly, he could not have alluded to Allégret, who had devoted the second chapter of his thesis to stressing the geometric significance of quaternions. Perhaps because of this view, we see Laisant consistently stressing in his work the geometric interpretation of quaternions[40]. This emphasis is visible from the definition of quaternions in terms of the biradial, that is, by introducing the quaternions as a quotient of vectors, even if this may not have been the best approach, as discussed in (Altmann 1989).

Laisant's remarks on commutativity and on the different status of geometry and analysis in France in that period are instructive and have some interest even outside the question of the acceptance of quaternions. His, certainly, are the perceptions of an active and well-connected member of the French mathematical community in the interesting years immediately after the Franco-Prussian War. As with other French mathematicians, Laisant accepted that opposition to quaternions in France had been strong. In a letter to the much younger geometer Maurice d'Ocagne (1862–1938), dated in 1893[41], Laisant still complained of the dominant position of analysis in contemporary French mathematics. He said that 'the great mathematicians of our time, who despise geometry and clarity, have never wished to understand equipollences except as an interpretation, which is absurd.' The gravity of such views is reinforced by an additional note in his letter: 'These remarks are only for you, as I wish to keep in good terms with all mathematicians, even those whose

[38] Laisant's notes in (Bellavitis 1874, p. ix).

[39] Laisant's notes in (Bellavitis 1874, p. ix).

[40] Laisant's notes in (Bellavitis 1874, pp. ix–x).

[41] Ch.-A. Laisant to M. d'Ocagne, March 9, 1893. *Archives de l'École Polytechnique*, Paris.

influence I regret most (as Jules Tannery).'[42] In another, earlier, letter, dated in the 1880s, Laisant made reference to a war being fought between those in favour and those against vectors in the mathematical Paris of the time.

Henri Brocard (1845–1922), a member of Laisant's wide circle of friends, published a review of Laisant's book in *Nouvelles Annales de Mathématiques* in July 1882. Brocard (Brocard 1882, p. 333) remarked that it was only after thirty years of Hamilton's publication that his work started to 'penetrate the mathematical French public.' He expressed the hope that through this book the method of quaternions might 'finish by triumphing over the prejudices that have, in France, retarded its development.'

The social side of mathematics: An international network of mathematical communication

While Laisant pursued his mathematical activities, he also developed an intense political career as a member of the French Chamber of Deputies; he adopted a more modern brand of socialism than Rodrigues's utopian speculations: these were the activities that he had described as 'numerous.' After Laisant left his interesting political career in 1893, he became more deeply concerned with the social side of the practice of mathematics, particularly with the development of a more integrated and coherent international community of mathematicians with much wider channels of communication. He participated in the organization of the first *International Congresses of Mathematicians* and founded important mathematical journals, such as *L'Enseignement Mathématique* and *L'Intermédiaire des Mathématiciens*, which were directed at the promotion of a more advanced mathematical education and to the exchange of mathematical knowledge between mathematicians of all levels. With his efforts to enhance the social side of the practice of mathematics, Laisant went much further than his social utopian colleagues would ever have dared to dream.

Rodrigues's descendants from 1840

The family tree of the descendants of Olinde Rodrigues down to the 1840s is given in Figure 2, Chapter 2, of this book. I will now consider some further sections of the tree, covering the 'after Rodrigues' period, which is considered in this chapter and extends from the second part of the nineteenth century to the early twentieth century.

This part of the family tree displays connections with a number of leaders of industry and with engineers. Both groups were particularly strong in the areas of banking, railways and other means of transportation, mines, and large technical state enterprises. All these areas were favored by the Saint-Simonians in their ambitious financial and engineering projects. The extended Rodrigues family has, as well, leading writers and scientists; many of them had a remarkable profile in public life. Here, I will concentrate on a section of this most interesting late-nineteenth and early-twentieth century family tree, with particular reference to the branches relevant to the exact sciences, leaving out at this stage others

[42] Jules Tannery was vice-director of the *École Normale Supérieure*. On the relations between the French normal and polytechnic schools, see (Zwerling 1990).

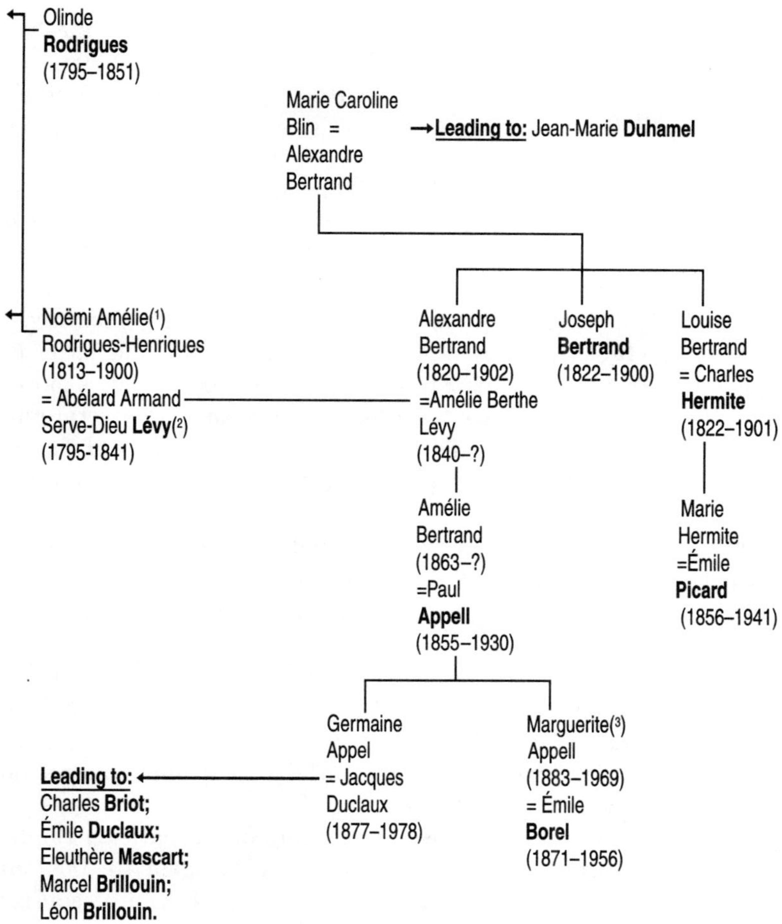

[1] In some documents Amélie Noëmi.

[2] In some documents Serve-Dieu Abéillard.

[3] Known as Camille Marbo.

Figure 1 . Olinde Rodrigues's family and its descendants after 1840.

which were also of interest. I will also leave out considerations on mathematical, rather than family, genealogy, which show the net to be even tighter.

At the lower right-hand side of the genealogical family tree of Rodrigues's descendants, given in Figure 2, Chapter 2, we find the name of Amélie Berthe Lévy, the daughter of Olinde Rodrigues's sister Noëmi Amélie Rodrigues-Henriques, who married mathematician and mineralogist Lévy. The career of the latter was sketched in Chapter 2; we would like to add that while in London, Lévy became acquainted with the English philosopher and social reformer Jeremy Bentham (1748–1832) and introduced him to Rodrigues.[43]

This niece of Olinde Rodrigues (see Figure 1 in this chapter[44]) married Alexander (1820–1902), the brother of Joseph Bertrand (1822–1900). The latter is a well-known mathematician and a leading figure in the French mathematics movement of the second half of the nineteenth century; his fame reached the general public and his name entered into Jules Verne's fiction. He became a member of the Academy in 1856 and later was elected its permanent secretary. He was also a professor at the *Collège de France.*

As is well known, his sister Louise Bertrand married Joseph's close friend, mathematician Charles Hermite (1822–1901). He was a devout Catholic and became a professor at the *École Polytechnique.* He was elected to the Academy in 1856, the same year as Bertrand.

The father of Alexandre, Joseph and Louise Bertrand, also called Alexandre (1795–1831)[45], came from a Catholic family. He and his close friend Jean-Marie Duhamel entered the *École Polytechnique* in 1814, the same year as Auguste Comte. Alexandre was a part of the group of founders of the Saint-Simonian periodical *Le Globe.*

Marie, daughter of Hermite, married the leading mathematician Émile Picard (1856–1941) in 1881. Picard became a member of the Academy in 1889 and later was also elected its permanent secretary. Meanwhile, Marie Hermite's cousin Amélie Bertrand married mathematician Paul Appell (1855–1930), who was born to a Catholic Alsatian family. Appell became professor and dean of the *Faculté des Sciences* of the University of Paris (1903–1920) and later its rector (1920–1925). He was elected to the Academy in 1892 and was active in government education councils.

In 1894–1906 France was aroused by the Dreyfus affair, an espionage matter in which anti-Semitic sentiment was agitated (Dreyfus being connected to the family of Jacques Hadamard's wife). Appell, together with Gaston Darboux, Paul Painlevé, Henri Poincaré, Jules Tannery, and other French mathematicians, joined in the

---

[43] I wish to thank Dr. C. Fuller, Bentham Project, for this reference.

[44] This tree has been compiled using information gathered at *Archives de l'Académie des Sciences, Archives Diplomatiques du Quai d'Orsay, Archives de l'École Polytechnique, Archives Nationales, Bibliothèque de l'Institut de France, and Bibliothèque Nationale de France,* all in Paris. Secondary sources include a variety of biographical and scientific dictionaries, personal memories and obituaries of some of the mathematicians whose names are mentioned in the family tree. The original sociologically oriented work of Charle (Charle 1987) and the works of Marbo (Marbo 1968) and Zerner (Zerner 1991) have also been consulted. Information in secondary sources is not always compatible and, whenever possible, has been contrasted with primary sources.

[45] As a physician, he wrote on somnambulism and contributed to demystifying animal magnetism.

efforts of the more radical Émile Zola and other French intellectuals to get the charges reversed[46].

Appell's daughter Marguerite[47] married the son of a Protestant pastor, mathematician Émile Borel (1871–1956). He became the leader of a world-class school on function theory and probability. He was designated professor at the *Sorbonne* in 1909, director of the *École Normale Supérieure* in 1910, and was elected to the Academy in 1921; in 1934 he became its president. In the First World War, Borel received the *Croix de Guerre*. In the period between the world wars he had an active and interesting public life. He was elected a member of the French Chamber of Deputies (1924–1936) and briefly, in 1925, minister for the navy. Borel's wife was a distinguished writer who adopted the pen name of Camille Marbo (the latter name being an abbreviation of Marguerite Borel). In some of her writings she discussed issues of racial and religious discrimination and feminism. The Borels had a country estate in Saint-Affrique, Mid-Pyrenees, where Borel was born. They used to holiday there with a group of academic friends. Under historian Charles Seignobos as their 'Capitan' (Marbo 1968, p. 176), members of this group developed strong personal bonds. The group included some leading exact scientists of the first quarter of the twentieth century, among them Marie Curie and her daughter Irène, Paul Langevin, Paul Painlevé, and Jean Perrin.

An elder sister of Marbo, Germaine Appell, married Jacques Duclaux (1877–1978), a chemical physicist and physician. He was the son of Émile Duclaux (1840–1904), a collaborator of Louis Pasteur and later a director of the Pasteur Institute. Through her, a link with another group of scientists was established: Émile's mother Mathilde Briot was the daughter of mathematician Charles Briot (1817–1882). Her sister Fanny married physicist and meteorologist Eleuthère (Éli) Mascart (1837–1908); their daughter married physicist Marcel Brillouin (1854–1948), the father of physicist Léon Brillouin (1889–1969).

Through intermarrying leading exact scientists, Olinde Rodrigues's extended family expanded to include an impressive array of research mathematicians, leaders of mathematics research schools, and influential academic administrators who played a relevant role in the development of the exact sciences in the second part of the nineteenth century and the early twentieth century in France. As already indicated, uniformity in religious faith was not the element binding this group together. Perhaps, the Saint-Simonian movement provided a sufficiently diverse scene that facilitated these personal encounters.

On a different but perhaps not unrelated subject, it should be mentioned that the Pereire family, closely linked to the Rodrigues clan and also connected in the past to the Saint-Simonian movement, occasionally expressed its patronage to science through donations to leading institutions such as the Academy (Zerner 1991, p. 309).

Younger members of the Rodrigues extended family were aware of Olinde Rodrigues's mathematical ability and of his work and made this view public. As Grattan-Guinness has indicated in Chapter 5, in his 1853 edition of Lagrange's

[46]In 1888 the German authorities accused Appell's brother Paul of espionage for France and imprisoned him. On the various attitudes of French mathematicians regarding the Dreyfus's affair, see (Rollet 2002)

[47]Referred to as Berthe in some secondary sources.

works (in particular, in notes to his *Mecánique Analytique* of 1788) Bertrand inserted notes crediting Rodrigues, quite rightly, with contributions to the 'principle des forces vives' and a few pages later quoted 'une Note remarquable d'Olinde Rodrigues.' Askey, in Chapter 6, showed in some detail that it was Hermite in 1865 who restored the name of the expression we know today as *Rodrigues's formula* to the name of the mathematician who first derived it. In Chapter 7, Tamm reminded us of a quotation by Bertrand in 1878 (Bertrand 1878) in which the mathematician stated that 'Lamé's elegant solution woke up an eminent spirit, for a long time oblivious to science. Olinde Rodrigues, in a note of a few pages, reminded his old fellow students, who by then had become great masters, that in the past, at the *Lycée* he led them all and that had he wished he could have become their equal.' Bertrand's evocation shows respect for Rodrigues's mathematical ingenuity, regretting that he had not pursued mathematics exclusively.

Final remarks

The transmission of quaternions to France is an interesting example of the complexity of the process of the diffusion of scientific knowledge between leading scientific nations with solid and well-established intellectual traditions. It is also an example of how patchy and irregular our knowledge of nineteenth-century mathematics still is.

There was institutional proximity between Laisant and Transon. In turn, the latter respected Rodrigues, his former companion both in the Saint-Simonian movement and in their jointly leaving it. There was as well some degree of shared mathematical interests between Laisant and Rodrigues. Their views on the need for society to change in order to offer better opportunities for all were different but not divergent. In spite of this, by the 1860s–1870s, when quaternions were transmitted from Great Britain to France, the importance of Rodrigues's work on the development of quaternions was overlooked. This suggests the existence of interesting fractures in mathematical thinking separating different periods, in spite of its protagonists having so much sympathy and ideas in common. The first of these periods is the one in which Rodrigues did his work. Allégret's marks a transition to Transon's period of explorations, culminating in its eventual acceptance in which members of Laisant's circle played a leading role.

The methodological divides between Rodrigues and the early group of French mathematicians interested in quaternions, up to Laisant's times, were perhaps too wide to be resolved due to the old French controversy between geometers and analysts, never far from the question of priority between analysis and synthesis.

It seems that the relationship between Rodrigues's early work and Hamilton's theory of quaternions began to be understood in France only at the turn of the century. In 1901 G. Koenigs, a conspicuous member of Laisant's circle of friends, opened his report on G. Combébiac's doctoral thesis *Calcul des triquaternions* with the following remarks: 'The formulae of Olinde Rodrigues are sufficient to provide evidence of the relations existing between the rotations around a fixed point and Hamilton's quaternions.'[48] Such a perceptive remark was largely ignored in the historical literature but clearly not within the French mathematical community of

[48] *Archives Nationals*, Paris, Séries $AJ^{16}/5538$ (1902–1904). The report is dated 17 October 1901; the thesis was defended on 22 October 1902.

the time, it should have had a more decisive impact on the evaluation of Rodrigues's results.

It is possible, however, that a tendency in the contemporaneous historical approach to science to associate new results with a cluster of eminent names may have contributed to obscure Rodrigues's contribution. Rodrigues did not have any students, but as we have just seen, members of his extended and influential family remembered his mathematical ability a quarter of a century after his death. It would appear that his name had not been completely forgotten in the main corridors of academic power in France during the second half of the nineteenth century.

In the time of Laisant, a reference to Rodrigues's work would have helped to keep national pride high, a point of some interest after the 1870 defeat. However, in Laisant's agenda for change in France's scientific community, his involuntary neglect of the Rodrigues episode in the history of quaternions was not without a positive result. Irrespective of his interest in quaternions, Laisant seemed keen on using two related theories, equipollences and quaternions, in order to illustrate his views on two fronts. The intellectual isolation of France, immediately before 1870, from ideas elaborated in Germany, Great Britain, and Italy was one of them. The need to attribute a wider space to geometrical theories and geometry in general within the French mathematical community's interests was the other. Laisant and his friends used both arguments to challenge the wisdom of the leadership of the French mathematical community of the time in an effort to promote change and, in their view, to open the subject to wider debate.

Bibliography

Allégret, A. (1862). *Essai sur le Calcul des Quaternions de M. W. Hamilton*, Thèse de doctorat ès-Sciences de la Faculté de Paris, No. d'ordre 242. E. Thunot et Cie (Leiber), Paris.

Altmann, S. L. (1989). Hamilton, Rodrigues, and the quaternion scandal. *Mathematics Magazine*, **62**, 291–308.

Arnoux, G. and N. N. Mousier (eds.) (1798). *Oeuvres Complètes de Condillac* (23 vols.), Vol. 23. Ch. Houel, Paris.

Baez, J. C. (2001). The octonions. *Bulletin of the American Mathematical Society*, **39**, 2, 145–205.

Bellavitis, G. (1835). Saggio di applicazioni di un nuovo metodo di geometria analitica (*Calcolo delle equipollenze*), *Annali Lombardo Veneto*, **5**, 244–250.

Bellavitis, G. (1838). Memoria sul metodo delle equipollenze. *Annali Lombardo Veneto*, **7** (1837), 243–261; **8** (1838) 17–37, 85–121.

Bellavitis, G. (1852). Saggio sull' algebra degli imaginarii, *Reale Istituto Veneto di scienze, lettere ed arti*, Memorie, **4**, 243–374.

Bellavitis, G. (1854). *Sposizione del metodo delle equipollenze.* Tipi Della Regio-ducal Camera, Modena. Also in *Memorie della Società Italiana delle Scienze residente in Modena*, **25**, 225–309. [This is the text translated by C. A. Laisant, Paris. 1873–1874.]

Bellavitis, G. (1858). Calcolo dei quaternioni de W. R. Hamilton e sua relazione col metodo delle equipollenze. *Memorie della Società Italiana delle Scienze residente in Modena*, 63 pp.

Bellavitis, G. (1873). (Transl. Charles-Ange Laisant.) Exposition de la méthode des equipollences, *Nouvelles Annales de Mathématiques*, **12**, 97–113, 145–161, 193–207, 241–257, 297–315, 501–519, 529–561; **13** (1874), 58–59, 138–152, 189–198, 220–236, 257–265.

Bellavitis, G. (1874). (Transl. Charles-Ange Laisant.) *Exposition de la méthode des équipollences.* Gauthier-Villars, Paris. [All the quotations in Roman numerals are from the Preface of the 1877 ed.]

Bellavitis, G. (1876). Sulle origini del metodo delle equipollenze. *Mémoires de l'Institut Impérial et Royal de Venice*, **29**, 449–491.

Benevolo, L, (1963). *Le origini dell'urbanistica moderna.* Biblioteca di Cultura Moderna, Bari.

Bertrand, J. (1878). Élogue de Gabriel Lamé, Ingénieur en Chef des Mines lu dans la séance publique annuelle du 28 janvier 1878. *Annales des Mines*, **7**, 13.

Brocard, H. (1882). Bibliographie (Introduction à la méthode des quaternions). *Nouvelles Annales de Mathématiques*, 3ème. série, **1**, 332–335.

Carnot, L. (1803). *Géométrie de position.* J. M. B. Duprat, Paris.

Charle, C. (1987). *Les élites de la République, 1880–1900*, Fayard, Paris.

De Morgan, A. (1849). *Trigonometry and double algebra.* London, Taylor, Walton & Maberly.

Fourcy, A. (1828). *Histoire de l'École Polytechnique.* École Polytechnique, Paris.

Freguglia, P. (1991). I fondamenti dell'algebra degli imaginari secondo Giusto Bellavitis, in *Giornate di storia della matematica* (M. Galuzzi, ed.), Editel, Rende, 143–168.

Gispert, H. (1991). La France Mathématique de France (1872–1914). *Cahiers d'Histoire et de Philosophie des Sciences, Nouvelle Série*, **34**, 1, 297–322.

Gispert, H. (2002). *Par la science, pour la patrie: L'Association Française pour l'Avancement des Sciences.* Les Presses Universitaires de Rennes.

Gray, J. J. (1980). Olinde Rodrigues' paper of 1840 on transformation groups. *Archive for History of Exact Sciences*, **21**, 375–385.

Halsted, G. B. (1897). Höuel. *The American Mathematical Monthly*, **4**, 4, 99–101.

Hamilton, W. R. (1853). *Lectures on quaternions.* Hodges & Smith, Dublin.

Hamilton, W. R. (1856). Memorandum respecting a new system of roots of unity. *Philosophical Magazine*, **12**, 446.

Hamilton, W. R. (1899). *Elements of Quaternions*, 2nd ed. (2 vols.). Longmans, Green and Co., London.

Hermite, C. (1865). Sur quelques développements en série de fonctions de plusieurs variables. *Ouvres de Charles Hermite*, **2**, (1918), Paris, 319–346.

Hoüel, J. (1855). *Applications de la méthode de M. Hamilton au calcul des perturbations de Jupiter.* Mallet-Bachelier, Paris.

Hoüel, J. (1858). *Tables de logarithmes a cinq décimales pour les nombres et les lignes trigonométriques; suivies des logarithmes d'addition et de soustraction ou logarithmes de Gauss et des diverses tables usuelles.* Mallet-Bachelier, Paris.

Hoüel, J. (1869*a*). Sur la méthode d'analyse géométrique de M. Bellavitis (Calcul des équipollences, par M. Bellavits). *Nouvelles Annales de Mathématiques*, 2 sér., **8**, 289–312, 337–357.

Hoüel, J. (1869*b*). *Sur le calcul des équipollences, méthode d'analyse géométrique de M. Bellavitis.* Gauthier-Villars, Paris, 46 pp.

Hoüel, J. (1874). *Éléments de la théorie des quaternions.* Gauthier-Villars, Paris, 296 pp.

Knight, I. (1968). *The geometrical spirit: The Abbé Condillac and the French Enlightenment.* Yale University Press, New Haven.

Knott, C. G. (1911). *Life and Scientific Work of Peter Guthrie Tait, Supplementing the two volumes of scientific papers published in 1898 and 1900.* Cambridge University Press.

Laisant, Charles-Ange (1874). Preface. In (Bellavitis 1874). [All the quotations are from the Preface of the 1887 ed.]

Laisant, Charles-Ange (1877). *Thèse*, No. de Ordre 297. Faculté des Sciences, Paris. 133 pp.

Laisant, Charles-Ange (1880). Giusto Bellavitis. *Bulletin des Sciences Mathématiques*, 2ème série, **4**, 343–348.

Laisant, Charles-Ange (1881). *Introduction à la méthode des quaternions.* Gauthier-Villars, Paris.

Marbo, C. (1968). *A travers deux siècles. Souvenirs et Rencontres.* Grasset, Paris.

Maxwell, J. C. (1890). *The Scientific Paper of James Clerck Maxwell* (W. D. Niven, ed.) (2 vols.). Cambridge University Press.

May, K. O. (1973). *Bibliography and research manual of the history of mathematics.* University of Toronto Press.

Möbius, A. F. (1856). Über eine neue Verwandtschaft zwitschen ebene Figuren. *Journal für die reine und angewandte Mathematik*, **52**, 218–228.

Mourey, C.-V. (1828). *La vraie théorie des quantités négative et des quantités prétendues imaginaires.* Bachelier, Paris. (Reprinted 1861 by Gauthier-Villars, Paris.)

Nový, L. (1973). *Origins of modern algebra.* Academia, Prague.

Ortiz, E. L. (1996). The nineteenth-century international mathematical community and its connection with those on the Iberian periphery, in *L'Europe Mathématique* (J. Gray, C. Goldstein, and J. Ritter, eds.), pp. 321–344. Edition de la Maison des Sciences, Paris.

Ortiz E. L. (1999). Quaternions abroad: some remarks on their impact in France. *Acta Historiae Naturalium Necnon Technicarum*, **3**, 295–302.

Ortiz E. L. (2003). Between algebra and analysis: The impact of the French *Idéologie* on Charles Babbage's work on functional equations. *The history of algebra in the nineteenth and twentieth centuries* (K. Parshall and J. Gray, eds.), Mathematical Sciences Research Institute Workshop, University of Berkeley, Berkeley, 1–38.

Prouhet, E. (1863). Bibliographie (Essai sur le calcul des quaternions de M. W. Hamilton). *Nouvelles Annales de Mathématiques*, 2ème série, **2**, 333–334.

Rollet, L. (2002), Un mathématicien dans l'affaire Dreyfus: Henri Poincaré. *Séminaire d'Histoire des Mathématiques—IHP* (13 février 2002), 1–13.

Smith, C. and M. Norton Wise (1989). *Energy and Empire: A biographical study of Lord Kelvin.* Cambridge University Press.

Tait, G. P. (1867). *An elementary treatise on quaternions.* Oxford University Press.

Tait, G. P. (1882). *Traité élémentaire des quaternions, Première Partie: Théorie, applications géométriques.* Gauthier-Villars, Paris.

Tait, G. P. (1884). *Traité élémentaire des quaternions, Seconde Partie: Géométrie des courbes et surfaces. Cinématique. Applications a la Physique.* Gauthier-Villars, Paris.

Tournaire, M. (1878). Notice Nécrologique sur Abel Transon. *Annales de Mines*, **14**, 433–449.

Transon, A. (1830). *De la religion saint-simonienne, cinq discours aux élèves de l'École Polytechnique. L'Organisateur*, Paris.

Transon, A. (1832). *Simple Écrit d'Abel Transon aux Saint-Simoniens.* Impr. de Éverat, Paris.

Transon, A. (1868*a*). De la séparation des racines. *Nouvelles Annales de Mathématiques*, **7**, 25–36, 57–67.

Transon, A. (1868*b*). Application de l'algèbre directive a la géométrie. *Nouvelles Annales de Mathématiques*, **7**, 145–157, 193–208, 241–264.

Transon, A. (1873). Sur un nouveau mode de construction des coniques. *Nouvelles Annales de Mathématiques*, **12**, 5–20.

Warren, J. (1828). *A treatise on the geometerical representation of the square roots of negative quantities.* Cambridge.

Zerner, M. (1991). Le règne de Joseph Bertrand (1874–1900). *Cahiers d'Histoire et de Philosophie des Sciences, Nouvelle Série*, **34**, 298–322.

Zwerling, C. S. (1990). *The emergence of the École Normale Supérieure as a center of scientific education in Nineteenth-Century France.* Garland, New York.

Archival sources

Archives de l'Académie des Sciences
Archives Diplomatiques du Quai d'Orsay
Archives de l'École Polytechnique
Archives Nationales
Bentham Project, University College London
Bibliothèque de l'Institut de France
Bibliothèque Nationale de France

Index

Titles in This Series